Problem Solving in Chemistry

Problem Solving in Chemistry

M. Selvaratnam and M. J. Frazer

A guide to solving numerical problems in general and physical chemistry at upper secondary and college level

Heinemann Educational Books

Heinemann Educational Books Ltd
22 Bedford Square, London WC1B 3HH

LONDON   EDINBURGH   MELBOURNE   AUCKLAND
HONG KONG   SINGAPORE   KUALA LUMPUR   NEW DELHI
IBADAN   NAIROBI   JOHANNESBURG
EXETER (NH)   KINGSTON   PORT OF SPAIN

British Library Cataloguing in Publication Data

Selvaratnam, M.
  Problem solving in chemistry
  1. Chemistry
  I. Title II. Frazer, M.J.
  540   QD31.2

  ISBN 0-435-65257-5

Printed in Great Britain by Spottiswoode Ballantyne Ltd.
Colchester and London

Preface

This book deals with numerical problems in elementary general and physical chemistry. It is suitable, as a supplementary text, for pre-university and first year university students.

Some of the features that differentiate this book from other books on problem solving are the following:

1 We place much greater emphasis on the methods and techniques that have to be applied for solving problems in the most effective manner. Throughout the book we describe how to analyse a problem, how to start, how to proceed, and how to obtain the correct solution. We present a 'five-step' procedure for solving problems and this procedure is used for the solution of all the numerical problems given in the book.

2 We point out how one should make use of problems, and their solutions, to reinforce the learning of the basic principles, laws, concepts, and applications of chemistry.

3 We also give some guidance on how to study effectively. A common difficulty of many students lies in not recognizing clearly which parts of a subject are fundamental and therefore have to be *remembered*, and which parts need not be remembered, where only familiarity is needed. We provide help to overcome this difficulty by giving only fundamental laws and equations in the text. These have to be understood thoroughly and remembered. Derived equations, consequences of laws, applications, elaborations, etc. are given as examples and exercises. These need not be memorized. The student is thereby helped by recognizing that only a few principles and equations need be memorized, and that these few principles and equations provide the foundation from which the student should start thinking when confronted with a problem.

We believe that the difficulties of many students in problem solving (and often in their studies too) are due to an improper method and technique. Students have difficulty in knowing how to start, where to start, and how to proceed with the analysis and solution of a problem. These aspects are discussed in Chapter 1 and are emphasized repeatedly in the worked examples throughout the book. A full understanding of Chapter 1 is not obtained easily. You are therefore strongly advised to study this chapter carefully several times, returning to it at different times during the course. The method and approach outlined in this chapter should become a natural part of your way of thinking.

Chapter 2 illustrates some of the necessary elementary mathematics, and it also emphasizes the importance of a proper choice of units.

You should always work in SI units. These units are now internationally accepted and a brief summary is included in Chapter 2. If you run into difficulties with a numerical problem in chemistry, always check that you have expressed all the quantities correctly in SI units.

Some fundamental concepts run throughout chemistry. These are illustrated, with worked examples, in Chapters 3 and 4. A thorough understanding and ability to use the subject matter in these chapters is essential for understanding the later chapters.

Problems from different areas of chemistry are then considered (Chapters 5 to 12). Since the main objective of this book is to help you to solve numerical problems in chemistry, there is not the same balanced coverage of topics you would find in a conventional text. Emphasis is always placed on the logical method of analysing problems. It is stressed repeatedly that to solve the problem only a few principles and equations are required.

For each topic there are three sections:

Firstly, there is a brief summary of the fundamental principles involved and wherever possible these are given in the form of equations defining the various physical properties. There are twenty-six such *defining equations* in this book and they are summarized in Appendix 3. With practice and understanding these equations *can and should be remembered*. This is not such a formidable task—go slowly—but remember that for successful problem solving in chemistry at this level it is essential to be able to recall and to understand these twenty-six equations. Every problem in this book can be solved by correctly applying one or more of these equations. Furthermore, it is our claim that most other numerical chemical problems at this level can be solved by a proper application of these equations.

Secondly, examples are given to illustrate the application of the fundamental principles or equations and the solutions are presented using the 'five-step' procedure. You should appreciate that the solutions are deliberately presented in complete detail. Once the 'five-step' procedure has been mastered there may not be the need to work consciously through the complete sequence. With practice the procedure will become automatic and you will only need to work through all the steps in the detail shown in the book when you are 'stuck' with a problem, or don't know what to do, or get a wrong answer.

Thirdly, some more problems are presented as student exercises. Numerical answers for these exercises will be found in Appendix 4 but detailed solutions are not given.

We believe that it is better for the student to analyse a few problems thoroughly rather than to 'muddle through' a large number. The number of examples and exercises has therefore been kept to a minimum.

To make the most effective use of this book you are advised to attempt working out the examples alone before reading through the solutions provided. Then compare your approach and answer with the solution provided. This procedure will help you to identify clearly the reasons for any failure. The necessity for mental effort on the part of the student cannot be over-emphasized. Just as physical effort is necessary for physical development, *mental effort is essential for intellectual development.*

Work through the examples and exercises, use the five-step procedure when you need to, and you will become a successful problem solver in chemistry. Good luck!

Acknowledgements

This book was initially written by one of us (M.S.). It was then tested on first year university and high school students in Sri Lanka; we thank the large number of students and teachers for their comments and suggestions that led to the improvement of the book. In particular our thanks are due to Professor S. G. Canagaratna (University of Peradeniya), Mr K. Balasanmugam (University of Peradeniya), and Mrs Hema Selvaratnam (Mahamaya College, Kandy) for their outstanding contributions. The book was then modified into its present form by our joint effort after useful comments and suggestions by Dr E. B. Robertson (University of Calgary, Canada), Mr M. Berry (Chislehurst and Sidcup Grammar School, UK), and Dr A. Ashmore (Royal Society of Chemistry, UK).

Contents

1 Problem solving— technique and importance

Solving problems is one of the most common difficulties for students of chemistry. This difficulty is often due to using an incorrect method and approach. Problem solving is made easier if the necessity for a logical and systematic approach is appreciated. In this chapter we discuss some of the most important aspects associated with this approach to problem solving in chemistry and then outline a five-step procedure for solving problems. Our discussion will be restricted to numerical problems.

1.1 Definition of the problem

Before you can solve a problem it is necessary to define clearly what the problem is. You should focus your attention on the objective; on what is required in the question. This statement appears obvious. But is it? Consider, for example, the following simple problem.

Example 1.1

3.00 g of phosphorus pentachloride (vapour) are heated in a closed 1.00 dm³ vessel at 300 °C. The degree of dissociation, according to the equation:

$$PCl_5(g) \rightleftharpoons PCl_3(g) + Cl_2(g)$$

is then 0.300. Calculate the density of the equilibrium mixture.

This problem has been tested on more than five hundred chemistry students at the University of Sri Lanka over the last several years, and also on university students in the US and UK. Over 75% of them could not solve it even though the solution is very simple. By definition,

$$\text{density} = \text{mass/volume}$$

$$= \frac{3.00 \text{ g}}{1.00 \text{ dm}^3}$$

$$= 3.00 \text{ g dm}^{-3}$$

The question arises: why did the majority of students find this problem difficult? Analysis of the answer scripts revealed the reason for the difficulty. The unsuccessful students did not define the problem. They started either with the equation for the dissociation

$$PCl_5(g) = PCl_3(g) + Cl_2(g)$$

or with the equation $pV = nRT$, and were attempting to fit the data into the equation without knowing 'where they were going'. The irrelevant data, which were provided deliberately, evidently caused confusion. This confusion could not have arisen had the students first defined the problem. Had the students first tried to define the problem by asking: 'What is it that has to be found out?' they would surely have solved it.

Question: What is the problem?
Answer: To determine density.
Question: What is the definition of density?

Answer: $$\text{density} = \frac{\text{mass}}{\text{volume}}$$

$$= \frac{3.00 \text{ g}}{1.00 \text{ dm}^3}$$

$$= 3.00 \text{ g dm}^{-3}$$

Many students are non-starters! This is often due to their not knowing *where* to start. As a general rule, we recommend that you start with the equation defining the required physical quantity. If this defining equation does not seem appropriate, then try another equation which relates the required physical quantity with at least some of the data provided.

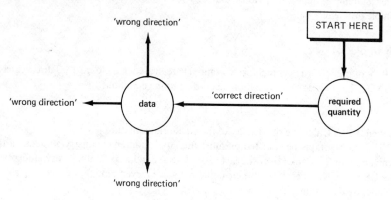

Fig. 1.1   Where to start?

Students sometimes start with the data and try to progress from there. This is not generally logical. If we start with the data, there is a greater chance of going in the 'wrong direction' than in the 'correct direction' (see Fig. 1.1). Do not therefore start with the data; instead start with a definition of the required quantity.

1.2 Clarification of the problem

It is often necessary to clarify a problem before being able to proceed. Clarification also makes the solution easier by helping to focus attention on the main aspects of the problem.

The word 'clarification' is used here to mean:

(a) making a precise and systematic *presentation* of all the given data, together with the physical quantity required
(b) *simplifying* the problem by making approximations and assumptions.

Two examples will illustrate problem clarification.

Example 1.2

Calculate the final pressure when a 1.00 dm³ flask of oxygen at 300 K and 1.00 atmosphere pressure is connected to a 5.00 dm³ flask of nitrogen at 300 K and 2.00 atmosphere pressure.

Making a precise and systematic presentation of the data
The problem is clearly presented in Fig. 1.2 which depicts the conditions in the initial state and in the final state.

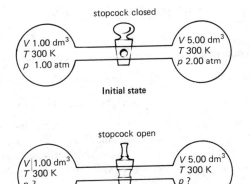

Fig. 1.2   A visual presentation of the data in Example 1.2

It is seen that all the relevant data, and the physical quantity required, have been arranged in a systematic and pictorial manner. We can then see the whole situation at a glance. Attention could then be focused on the problem without distraction. It is easier to think while looking at a figure rather than while reading the words given in a problem.

In some problems the difficulty arises because of our inability to keep in our minds the different aspects of the problem at the same time. Presenting the different aspects of a problem together, as in Fig. 1.2, helps to reduce this difficulty.

The problem is to find the final pressure p. The figure of the final state indicates that p is the sum of the partial pressures of oxygen and nitrogen and this helps in devising the method of solution*.

---

* The objective in this chapter is to illustrate the *method* of analysing and solving problems. The complete solution is therefore not given here.

Example 1.3

Calculate the concentration of hydrogen ions in a 0.10 M* solution of ortho-phosphoric acid in water which dissociates, in three steps, according to the equations

$$H_3PO_4(aq) \rightleftharpoons H^+(aq) + H_2PO_4^-(aq) \tag{1}$$

$$H_2PO_4^-(aq) \rightleftharpoons H^+(aq) + HPO_4^{2-}(aq) \tag{2}$$

$$HPO_4^{2-}(aq) \rightleftharpoons H^+(aq) + PO_4^{3-}(aq) \tag{3}$$

The acid dissociation constants for the first, second, and third dissociations are 1×10^{-2} mol dm^{-3}, 1×10^{-7} mol dm^{-3}, and 1×10^{-13} mol dm^{-3} respectively.

Simplifying the problem by making approximations

This requires an understanding of the subject matter (for Example 1.3, see Chapter 8) and sound judgement which comes with practice and experience. It is seen from the three dissociations that the concentration of hydrogen ions present in the solution, $[H^+]$, is given by

$$[H^+] = [H^+]_1 + [H^+]_2 + [H^+]_3 \tag{4}$$

where

$[H^+]_1$ = concentration of hydrogen ions produced by dissociation (1)

$[H^+]_2$ = concentration of hydrogen ions produced by dissociation (2)

$[H^+]_3$ = concentration of hydrogen ions produced by dissociation (3)

To calculate $[H^+]$ by making use of equation (4) (i.e. by calculating $[H^+]_1$, $[H^+]_2$, and $[H^+]_3$) is not only difficult but also quite unnecessary. This is because it can be assumed by examining the value of the given acid dissociation constants that $[H^+]_2$ and $[H^+]_3$ are negligibly small compared with $[H^+]_1$; we can therefore simplify equation (4) and write

$$[H^+] = [H^+]_1 \tag{5}$$

To calculate the hydrogen ion concentration of the solution, $[H^+]$, we have therefore to consider only equation (1).

How do we know that $[H^+]_2$ and $[H^+]_3$ would be negligibly small compared with $[H^+]_1$? This information is provided by the values given for the dissociation constants. The first dissociation constant (1×10^{-2} mol dm^{-3}) is seen to be 10^5 times (i.e. 100 000 times) larger than the second dissociation constant (1×10^{-7} mol dm^{-3}), and 10^{11} times larger than the third dissociation constant. Since the dissociation constant is a measure of the amount of dissociation we can assume that $[H^+]_2$ and $[H^+]_3$ are negligibly small compared with $[H^+]_1$.

1.3 Analysis of the problem

We have already emphasized that you should start the analysis of a problem with an equation involving the physical quantity required. We call this the key equation. You

---

* The symbol M is used to represent concentrations in mol dm^{-3}. (See footnote on page 62.)

should then proceed step by step with equations for each of the unknown quantities involved in the key equation. We illustrate this method with the help of an example.

Example 1.4

A vessel contains 3.00 g of a gas of molar mass 32.0 g mol^{-1}. The pressure exerted by the gas is 2.52 × 10^5 Pa at a temperature 300 K. Calculate the density of the gas given that the gas constant $R = 8.314$ J K^{-1} mol^{-1}.

*Analysis**

We are asked to calculate the density of the gas. We should therefore *start* with an equation defining density.

Density (ρ) is defined as the mass (m) per unit volume (V). That is

$$\rho = \frac{m}{V} \tag{1}$$

This equation shows that to determine density we must know both m and V. Now look at the data to see whether these quantities have been given. You will see that m is given, but the second quantity required, V, is not given. We must therefore try to calculate V from the given data.

To calculate V from the data given, we have to think of an equation which shows the relationship between V and the data. We try the ideal gas equation $pV = nRT$ (see **Equation 5.1**) which on rearrangement gives

$$V = \frac{nRT}{p} \tag{2}$$

Equation (2) shows that to calculate V we must know n, R, T, and p. Now look again at the data; values of R, T, and p are known. n, however, is not given.

To calculate V using equation (2) we require n. The value of n can be calculated from the data by making use of the following relationship between molar mass M, mass m, and number of moles n (see **Equation 3.3**).

$$n = \frac{m}{M} \tag{3}$$

m and M are given in the data. It follows therefore that we could calculate n, and therefore V (by equation 2), and therefore the required quantity ρ (by equation 1).

Note clearly the method of analysis. There is a step by step analysis starting with the key equation for the quantity required. At each step we look into the data with a definite objective; to see whether the quantities required for the calculation are given. Irrelevant data will then not even be 'seen'; any further data necessary for the calculation can also be predicted once the analysis is complete.

In Example 1.4, three stages were necessary for the solution. The first stage was to define the required quantity by an equation (the key equation) and this was followed by two further stages each introducing a new equation. In all, three equations were necessary for the solution.

---

* We are interested here only in the *method* of analysis and so the derivation of equations and the complete solution are not given. You can refer to Chaper 5 for explanation.

To simplify the calculation, the three equations (1), (2), and (3) should be combined together into one equation, before inserting the numerical values in order to obtain the answer. If we substitute for V in equation (1) by equation (2) we get

$$\rho = \frac{m}{V}$$

$$= \frac{m}{(nRT/p)}$$

$$= \frac{mp}{nRT} \tag{4}$$

If we then substitute for n in equation (4) by equation (3) we obtain

$$\rho = \frac{mp}{(m/M)RT}$$

$$= \frac{pM}{RT} \tag{5}$$

By substitution in equation (5), the density could be calculated since values for all the physical quantities on the right side of the equation are given in the data.

To be successful in problem solving flexibility of thought is necessary. In Example 1.4 the key equation (that is the equation used as a basis for solving the problem because it linked the required physical quantity, density, with one of the quantities given in the data) was the same as the defining equation for density. But this will not always be so and you should be prepared sometimes to use as the key equation one different from the defining equation. The next example is an illustration.

Example 1.5

Calculate the relative molecular mass ('molecular weight') of a substance if 0.0100 mol of it has a mass 1.00 g.

The definition of relative molecular mass is in terms of the ^{12}C isotope of carbon (the definition is given in section 3.1). The given data do not relate directly to this definition. We must therefore look for a different equation to be the key equation. We would need to remember (see Example 3.4) that the relative molecular mass M_r is numerically equal to the molar mass M when expressed in grams. To calculate M from the data given, we must use an equation which shows the relationship between M and the data. The required equation (**Equation 3.3**) is

$$M = \frac{m}{n}$$

where n is the moles, m is the mass, and M is the molar mass. Since n and m are given in the data, M could be calculated by making use of the above equation. Hence M_r could be obtained.

1.4 The use of equations

Many problems in chemistry involve calculations. Since calculations generally require equations, it is clear that you must understand, remember, and be able to manipulate equations.

Suppose you are asked in a problem to calculate a particular physical quantity. To do so, you should start with an equation (the key equation) involving that quantity. There may be, however, several equations involving the physical quantity—one of them will be the defining equation. How do you then decide which equation to select as the key equation? For help in making this decision look at the data given. The equation selected should show the relationship between the physical quantity required (M, in Example 1.5) and the physical quantities either given in the data (n and m, in Example 1.5), or calculable from the data.

Students sometimes work problems starting with quantitative statements such as 'one mole of gas occupies 22.4 dm^3 at 273 K and one atmosphere pressure'. This statement follows from the equation $pV = nRT$ (see **Equation 5.1**). There are several advantages in working problems using equations rather than using statements corresponding to the equation. Calculations using equations are more precise, less time-consuming, and less error-prone than calculations based on statements. A tighter check on the calculation is also possible; units too are directly involved in the calculation.

In any quantitative study of science, equations are necessary. They cannot be avoided for long. It is better for you to *learn how to use equations for calculations* rather than 'dodge' them by the use of equivalent statements.

1.5 Calculations

The solution of a numerical problem always involves a calculation after inserting values for the physical quantities in an equation.

For calculating the answer, certain *mathematical operations* (generally simple arithmetical operations) have to be carried out. The mathematics necessary is explained in Chapter 2.

When giving values for the physical quantities remember always to insert the *units* along with the numerical values. For instance, in Example 1.1 the density (ρ) should be calculated as

$$\rho = \frac{\text{mass}}{\text{volume}}$$

$$= \frac{3.00 \text{ g}}{1.00 \text{ dm}^3} = 3.00 \text{ g dm}^{-3}$$

and not as

$$\rho = \frac{3.00}{1.00} = 3.00$$

The units for the various physical quantities involved in the calculation should be consistent with each other. A short account on units, and the advantages of inserting the units along with the numerical value for any physical quantity, is given in Section 2.4.

How accurately should a calculation be carried out? This depends on the accuracy with which the data are given. As a general rule, the accuracy of a calculation should not exceed the accuracy of the least accurate item of data. The answer obtained should also not be expressed more accurately than that warranted by the data. For example, the answer obtained in Example 1.1 should be given as 3.00 g dm^{-3} and not 3.000 g dm^{-3} (if the answer is given as 3.000 g dm^{-3}, the accuracy claimed is 1 in 3000!).

1.6 Reviewing and checking the solution

Many errors in problem solving are due to carelessness. Therefore, you should always finally read through and review the problem statement and the solution. Make sure of the following:

(a) that you have answered the question actually asked in the problem statement
(b) that the answer obtained is of the correct order of magnitude (this could be checked by doing an approximate calculation and so any error in the placing of the decimal point in the answer, for example, would then be spotted)
(c) that the answer obtained is not unreasonable (for example, suppose that your calculation leads to an answer of 0.50 for the relative atomic mass of an element. This is an unreasonable answer because from its definition it is clear that the relative atomic mass can never be less than 1—the calculation has therefore to be checked)
(d) that any assumptions or approximations made during the calculation are justifiable
(e) that the units of the answer are correct
(f) that the number of significant figures in the answer is appropriate
(g) that you identify and understand clearly all the principles involved in the calculation (this point is emphasized in section 1.7).

1.7 Importance of problem solving

It is worth learning how to solve chemical problems for three reasons.

(a) If you ever become a professional chemist, the difference between a right and wrong answer might be a matter of life and death, or an investment of a million pounds might be at stake. At school level, you will want to obtain as many correct answers as possible in tests and examinations.
(b) By learning the skills of problem solving in chemistry you will contribute to your intellectual development and you will be more able to solve problems in other fields. The general principles of problem solving are universal even though we are concentrating on numerical chemical problems in this book.
(c) You will gain a better understanding of more chemistry. When you have succeeded with a problem, don't sit back in satisfaction but always go back and see what more you can learn. Review the concepts, laws, and principles which you used in your solution. You will be surprised how often there is something new to learn.

Example 1.6 will show you how useful information might be extracted from a problem.

Example 1.6

In a Victor Meyer's apparatus 0.0500 g of an organic liquid displaces, on vaporization, 22.0 cm^3 of air at 298 K and a pressure of 1.0×10^5 Pa. Calculate the molar mass of the liquid.

Information

This problem outlines Victor Meyer's method for the determination of molar mass. Even if we had been ignorant of the method we can, by deduction from the data, get some idea of the method, its scope, and its limitations.

The amounts of liquid used and gas collected give some information about the size of the apparatus. The necessity for vaporization suggests that only volatile liquids that do not decompose on heating can be used. Some heating device is also obviously necessary (for vaporization of the liquid). The calculation of the molar mass M from the data requires (see Chapter 5) the use of the equation $pV = (m/M)RT$. This indicates that the method depends basically on finding the volume (V) of a known mass (m) of gas at a known pressure (p) and known temperature (T). Some of the experimental precautions can therefore be deduced. For example, the temperature and the pressure have to be known and kept constant because the value of M depends on these factors.

1.8 Summary—the five-step procedure for solving numerical problems in chemistry

In this section we bring together all the ideas developed in the earlier sections and show you a simple procedure with five steps which forms the basis for solving all the examples and exercises given in this book.

In a numerical problem you have usually to calculate the value of a particular physical quantity from values of some other physical quantities given in the problem statement. To do so the simplest procedure, we suggest, would generally be first to *derive an equation which shows how the required physical quantity is related to the physical quantities given in the data*. The crucial question therefore is: 'What is the best procedure for deriving this equation?' A simple procedure, in operational form, for deriving such an equation is outlined below in steps 1–3; the calculation, checking, and learning aspects are then summarized in steps 4–5.

You are advised to read this section several times during your study of this book. The five-step procedure will become increasingly clear to you as you practise with the examples and exercises in later chapters.

Step 1 Clarify and define the problem

Objective

To obtain an overall coordinated view of the entire problem and to identify what should be done operationally in order to solve the problem.

Operations

(a) Read quickly through the problem statement.
(b) Give symbols for each of the physical quantities stated in the problem and collect and arrange all the data and the physical quantity required in a systematic and convenient manner (e.g. in a picture, graph, or table) so that a coordinated view of the entire problem may be seen at a glance.
(c) Focus attention on what has to be found out and establish clearly what must be done operationally in order to solve the problem. For this purpose a rephrasing of the problem statement may be helpful.

Step 2 Select the key equation

Objective

To select the most appropriate equation (called the key equation) with which to start the solution of the problem.

Operations

(a) Write down the defining equation* for the required (i.e. unknown) physical quantity.
(b) Write down any other equations which relate some of the given data to the required physical quantity.
(c) Check whether there are any limitations under the conditions stated in the problem on the application of any of the equations from (a) and (b). For the equations valid under the stated conditions proceed to (d).
(d) Select as the key equation a valid equation which relates the required physical quantity with the largest number of items of the data given in the problem. (Very often you will find that the defining equation for the required physical quantity is suitable as the key equation.)

Step 3 Derive the equation for the calculation

Objective

To derive an equation which can be used for the calculation of the required physical quantity. In this equation there should be only one unknown quantity—the required physical quantity.

Operations

(a) Rearrange the key equation (if this is necessary) so that only the required physical quantity is present on the left side of the equation; all the other quantities would then be on the right side.
(b) Identify any physical constants in the key equation. These are known quantities since their values will be obtained from tables of physical constants (e.g. relative atomic masses, Avogadro constant, Faraday constant, gas constant).
(c) Replace, step by step, each of the unknown quantities on the right side of the key equation with known quantities. To do so select the most appropriate

---

* For a few problems there may not be a defining equation. The 26 defining equations covering nearly all the chemistry in this book are given in Appendix 3.

equation for each unknown quantity and combine it with the key equation. The procedure for the selection of the most appropriate equation is exactly similar to that for the selection of the key equation; we have to perform again operations (a), (b), (c), and (d) of Step 2.

Repeat the procedure until there are no unknown physical quantities on the right side of the equation.

(d) If you are unable to derive the equation for the calculation try to do one or more of the following:

  (i) make use of other equations as the key equation

  (ii) make use of other equations for replacing the unknown quantities in the key equation

  (iii) make some simplifying assumptions or approximations.

Step 4 Collect the data, check the units, and calculate

Objective

To write down together the values for all the physical quantities required for the calculation (using the equation for the calculation), to check these values, and to calculate.

Operations

(a) Refer to the list of values of all the physical quantities which you wrote down from the given data in Step 1 (b) and check that the units are consistent. Make any changes necessary in the values so that all quantities are in consistent units.

(b) Insert these values (with units) into the equation for the calculation and calculate the answer. The units for the answer should also be deduced at the same time.

Step 5 Review, check, and learn from the solution

Objective

To check that the solution is complete and correct, and to review the solution so as to reinforce your learning of problem solving skills and of the principles of chemistry.

Operations

(a) Read through the problem statement and make sure that the answer obtained is for the question asked and also that all parts of the problem have been solved.

(b) Estimate approximately the order of magnitude of the answer and compare it with the answer obtained. This is to ensure that no gross errors (e.g. error in the placing of the decimal point) have been made in the calculation.

(c) Check the answer to see that it is of the correct sign, that it is not unreasonable, and that the number of significant figures is appropriate.

(d) Check that the answer is given in the correct units.

(e) Make use of the answer obtained to check that any approximations or assumptions made during the calculation are justifiable.

(f) Review and analyse the solution in order (i) to obtain an overall view of how the problem was solved, (ii) to identify any steps that caused difficulty and the reasons for such difficulty, and (iii) to reinforce your problem solving skills.

(g) Identify the basic equations and principles that were required for the solution and make sure that you can recall, understand, and apply them.

(h) Recognize and remember any useful information provided by the problem and its solution.

2 | Elementary mathematics

Some students find problem solving difficult because of the mathematics involved. This chapter outlines briefly some of the necessary mathematics, and illustrates it with some simple examples from chemistry. Practise the mathematics in this chapter and refer back to it when in difficulty. More mathematics is introduced, along with chemistry, in some of the later chapters.

2.1 Numbers

Small and large numbers are generally given in terms of exponents (index or power) of the number 10. For example

$$1000 = 1.000 \times 10^3$$

$$1200 = 1.200 \times 10^3$$

$$0.001\,200 = 1.200/1000 = 1.200/10^3 = 1.200 \times 10^{-3}$$

The general rules for dealing with exponential numbers are summarized in Table 2.1. The rules are true whatever the value of the number a, n, and m.

Table 2.1   General rules for exponential numbers

	Rule	Example
1	$a^n = a \times a \times a \times \dots$ (n times)	$10^4 = 10 \times 10 \times 10 \times 10 = 10\,000$
2	$a^{-n} = 1/a^n$	$10^{-4} = 1/10^4 = 0.0001$
3	$a^0 = 1$	$10^0 = 1$
4	$a^n \times a^m = a^{n+m}$	$10^3 \times 10^4 = 10^{3+4} = 10^7$
5	$a^n/a^m = a^{n-m}$	$10^3/10^6 = 10^{3-6} = 10^{-3}$
6	$(a^n)^m = a^{nm}$	$(10^3)^4 = 10^{3\times4} = 10^{12}$
7	$(a \times b)^n = a^n \times b^n$	$(10 \times 4)^3 = 10^3 \times 4^3$
8	$a^{1/n} = \sqrt[n]{a}$	$(27)^{1/3} = \sqrt[3]{27} = 3$

Example 2.1

The number of molecules in one mole of a compound is 6.022×10^{23}.

(a) Express this number in the normal manner.
(b) Calculate the mass of one molecule of H_2O if the mass of one mole of H_2O is 18.02 grams.

Solution

(a) Application of Rule 1 given in section 2.1 shows that

$$6.022 \times 10^{23} = 602\ 200\ 000\ 000\ 000\ 000\ 000\ 000$$

(b) From data

mass of one mole (i.e. 6.022×10^{23} molecules) of $H_2O = 18.02$ g

$$\therefore \qquad \text{mass of 1 molecule} = \frac{18.02\ \text{g}}{6.022 \times 10^{23}}$$

$$= \frac{18.02 \times 10^{-23}\ \text{g}}{6.022} \qquad \text{(by Rule 2 above)}$$

$$= 2.992 \times 10^{-23}\ \text{g}$$

Example 2.2

Suppose that the Avogadro constant (6.022×10^{23}) of dollars is distributed equally among the 4000 million inhabitants of the world. How many million dollars will each person then receive (1 million $= 10^6$)?

Solution

$$4000\ \text{million} = 4000 \times 10^6 = 4.0 \times 10^3 \times 10^6$$

$$= 4.0 \times 10^9 \qquad \text{(by Rule 4)}$$

4.0×10^9 people receive 6.022×10^{23} dollars

$$\therefore \qquad \text{amount received by 1 person} = \frac{6.022 \times 10^{23}}{4.0 \times 10^9}\ \text{dollars}$$

$$= \frac{6.022}{4.0} \times 10^{14}\ \text{dollars} \qquad \text{(by Rule 5)}$$

$$= 1.5 \times 10^{14}\ \text{dollars}$$

$$= 150\ 000\ 000\ \text{million dollars}$$

Exercise 2.1

(a) Is 10^{-7} a positive number or a negative number?

(b) Arrange the following numbers in decreasing order, i.e. start with the largest number:

$$5 \times 10^3,\ 10^4,\ 10^3,\ 10^0,\ 1.5,\ 0,\ 10^{-7},\ 8 \times 10^{-7},\ 10^{-6}$$

(c) Express each of the following numbers as a power of 10 (e.g. $100 = 1.00 \times 10^2$):

    (i) 10 000   (ii) 0.001   (iii) $100^{1.5}$

(d) Find the value of x in each of the following equations:

    (i) $96\ 500 = 9.6500 \times 10^x$

    (ii) $0.00123 = 1.23 \times 10^x$

    (iii) $12.3 \times 10^{-3} = 1.23 \times 10^x$

    (iv) $1.2 \times 10^3 x = 1$

(e) $10^{-2} \times 10^{-7} \times 10^5 =$

(f) $\dfrac{10^{-2}}{5 \times 10^{-8} \times 10^3} =$

(g) $\dfrac{0.0015 \times 10^{23}}{2 \times 10^{-2}} =$

(h)   By how many times is 5×10^{-4} greater than 2.5×10^{-7}?
(i) $1.000 \times 10^2 + 1.0 \times 10^{-1} =$
(j) $1.00 \times 10^{-5} - 1.00 \times 10^{-7} =$
(k) $(4 \times 10^4)^{1/2} =$
(l) $(2^0 \times 10^{-2})^2 =$
(m) $(3 \times 10^{-7})^3 =$
(n) $(8 \times 10^{-9})^{1/3} =$
(o)   If $r^3 = 9 \times 10^{24}$, calculate r.
(p)   If $1\ m = 10^3\ mm$, find the relationship between $1\ m^3$ and $1\ mm^3$.

(Check your solutions with those given on page 219.)

Exercise 2.2

The volume (V) of a sphere is given by the equation $V = (4/3)\pi r^3$ where r is the radius and $\pi = 3.14$. By assuming that the hydrogen atom is a sphere, calculate its volume. The radius of the hydrogen atom $= 3.7 \times 10^{-11}\ m$.

Exercise 2.3

An atom is built up of a minute positively-charged spherical nucleus around which negatively-charged particles, called electrons, move at very fast speeds. Calculate the radius of the nucleus of a hydrogen atom given that the volume of the nucleus is 1.0×10^{12} times smaller than that of the atom. The volume of a hydrogen atom is $2.0 \times 10^{-31}\ m^3$.

Exercise 2.4

A mole of particles contains the Avogadro constant (6.022×10^{23}) of particles. A mole of electrons has a charge of 96 490 coulombs. Calculate the charge on an electron.

Exercise 2.5

Light travels at a speed of 186 000 miles per second. Even then it takes about $4\frac{1}{2}$ years for light from the nearest star (known as Alpha Centauri) to reach the earth. What is the distance in kilometres (km) of Alpha Centauri from the earth? $1\ km = 0.62$ mile.

2.2 Mathematical equations

Importance of equations

Equations are important because they summarize quantitative information in a precise manner. They show quantitatively how a particular physical property depends on other physical properties.

As an example, consider the ideal gas equation $pV = nRT$ where R is a constant. This equation shows quantitatively how the different variables (the four variables p, V, n, and T) are related to one another. A clear understanding of this equation corresponds to a good understanding of the physical behaviour of gases. An analysis of this equation so as to reveal the information it contains is given in Chapter 5.

You should learn how to analyse equations so as to extract from them the information they contain. We shall illustrate the method of analysing equations when we come across specific equations in later chapters.

Mathematical meaning of some terms

To work through a problem successfully we must have a clear understanding of the terms and statements given in the problem, and also of the necessary laws. We explain here terms that sometimes cause confusion, and also express the meaning of each of these terms in equation form.

Per

The word 'per' means 'divided by'.

For example, density is defined as the mass per unit volume. This means that

$$\text{density} = \frac{\text{mass}}{\text{volume}}$$

Similarly, concentration is defined as the amount of substance per unit volume. This means that

$$\text{concentration} = \frac{\text{amount of substance}}{\text{volume}}$$

Fraction

The word 'fraction' refers to an amount divided by the total amount.

For example, the fraction of a substance A in a sample is given by the equation

$$\text{fraction of A} = \frac{\text{amount of A in sample}}{\text{total amount of sample}} \qquad \textbf{Equation 2.1}$$

Similarly the fraction (or degree) of dissociation α is given by the equation

$$\alpha = \frac{\text{amount dissociated}}{\text{total amount initially}} \qquad \textbf{Equation 2.2}$$

Percentage (%)

Percentage is defined by the equation

$$\text{percentage} = \text{fraction} \times 100 \qquad \textbf{Equation 2.3}$$

For example

$$\text{\% of A in a sample} = \frac{\text{amount of A in sample}}{\text{total amount of sample}} \times 100$$

$$\% \text{ dissociation} = \frac{\text{amount dissociated}}{\text{total amount initially}} \times 100$$

Directly proportional and inversely proportional

Consider two physical quantities x and y.

If x is said to be *directly proportional* to y, it means that the equation connecting x and y is

$$x = ky$$

where k is a constant.

On the other hand, if x is said to be *inversely proportional* to y, it means that the equation is

$$x = k\frac{1}{y}$$

Mathematical meaning of statements of laws

Problem solving often involves calculations. Calculations are done better with equations than with statements. Calculations using equations are more precise, are often easier, and are less prone to error than calculations using statements. Statements are more difficult to manipulate than equations. Precision of thought is encouraged by equations.

Any precise statement (e.g. a statement of a law) can be converted into an equation. You should learn the skill of converting a statement into an equation. This skill is necessary for the solution of many problems. Two examples are now given.

Example 2.3

Boyle's law states that, at constant temperature, the pressure of a fixed amount of gas is inversely proportional to its volume. Convert this statement into an equation.

Solution

To convert a statement into an equation the first step always is to give symbols for the given physical quantities. Let p be the pressure and V the volume. The statement 'pressure (p) is inversely proportional to volume (V)' can be represented in mathematical form as

$$p \propto 1/V \tag{1}$$

Whenever there is a proportionality sign in a relation we can replace it by an equal sign and a constant. Therefore equation (1) can be rewritten as

$$p = k/V$$

where k, which is a constant, is often known as the constant of proportionality.

Example 2.4

Raoult's law states that the partial vapour pressure of a component in a solution is directly proportional to the mole fraction of that component. Express this law as an equation.

Solution

The first step is to give symbols. Let the component be denoted by A, its partial pressure by p_A, and its mole fraction by x_A.

Raoult's law statement could then be given as

$$p_A \propto x_A$$

$$\therefore \qquad p_A = kx_A \qquad (1)$$

where k is the constant of proportionality.

The physical meaning of a constant term in an equation can often be obtained by analysing the equation. Look at equation (1). When $x_A = 1$ it is seen that $k = p_A$. What does $x_A = 1$ mean? From the definition of mole fraction (see **Equation 4.4**) it follows that $x_A = 1$ corresponds to pure A. k is therefore the vapour pressure of pure A (which let us denote by p_A^0). That is, $k = p_A^0$. Equation (1) can therefore be rewritten as

$$p_A = x_A\, p_A^0$$

Note that the above equation gives more information than is apparent from the *statement* of Raoult's law, or from equation (1). Raoult's law is discussed on page 86.

Exercise 2.6

Charles' law states that, at constant pressure, the volume of a fixed mass of gas is directly proportional to its temperature (T). Give this statement as an equation.

Exercise 2.7

Graham's law states that the rate of effusion (v) of a gas is inversely proportional to the square root of its density (ρ). Convert this statement into an equation.

By making use of Graham's law derive an equation which can be used to compare the rates of effusion of two gases having densities ρ_1 and ρ_2.

2.3 Logarithms

We can represent any number y as an exponent, or power, of another number A as follows

$$y = A^x$$

x is then known as the logarithm of the number y to the *base A*. This definition can be given in equation form as follows

$$x = \log_A y$$

y is then known as the antilogarithm of x.

Look carefully at the two equations given above. Note and remember how the logarithm (x) and the antilogarithm (y) are related.

Two bases for logarithms are commonly used—the base 10 and the base e. When no base is specified for a logarithm it should be assumed that the base is 10. The number e is the sum of the following series of terms

$$e = 1 + \frac{1}{1} + \frac{1}{1 \times 2} + \frac{1}{1 \times 2 \times 3} + \frac{1}{1 \times 2 \times 3 \times 4} + \ldots \text{ to inifinity}$$

e has a value 2.718.... Note that even though an infinite number of terms are added, the value of e is finite, i.e. 2.718.... The number e is important because it occurs in many important equations in the physical sciences.

Logarithms to the base e are known as natural logarithms or Napierian logarithms. For any number y the relationship between logarithms to the base e (denoted $\log_e y$ or $\ln y$) and logarithms to the base 10 (denoted by $\log_{10} y$) can be shown to be given by the equation

$$\ln y = \log_e y = \log_e 10 \times \log_{10} y$$

$$= \log_{2.718} 10 \times \log_{10} y$$

$$= 2.303 \times \log_{10} y \qquad \textbf{Equation 2.4}$$

Natural logarithms can therefore be converted to logarithms to the base 10; the conversion factor 2.303 should be remembered.

Table 2.2 gives the logarithms ($\log_{10}$) of some numbers. The logarithm of the number in the first row is given in the second row. The exponential form of each number in the first row is also indicated within brackets. It is clear from the definition of antilogarithms that the antilogarithm of each number in the second row is given by the number in the first row.

Table 2.2   Logarithms of some numbers

Number	0.00001 $(=10^{-5})$	0.01 $(=10^{-2})$	0.1 $(=10^{-1})$	1 $(=10^0)$	10 $(=10^1)$	100 $(=10^2)$	100 000 $(=10^5)$
Logarithm (to base 10)	-5 (or $\bar{5}$)	-2 $(\bar{2})$	-1 $(\bar{1})$	0	1	2	5

Since the logarithm is an *exponent* it follows that the rules for operating logarithms follow directly from the rules for dealing with exponential numbers (given in section 2.1).

Some of the useful rules which you should remember are

$$\log (A \times B) = \log A + \log B$$
$$\log (A/B) = \log A - \log B \qquad \textbf{Equation 2.5}$$
$$\log a^n = n \log A$$

From the definition of antilogarithms it also follows that

$$\text{antilog} (A + B) = \text{antilog } A \times \text{antilog } B \qquad \textbf{Equation 2.6}$$

Example 2.5

Find $\log_{10}$ of (a) 25.00 (b) 2.500×10^8 (c) 0.02500 (Log tables are given at the end of the book.)

Solution

(a) To find the logarithm to the base 10 of any number, it is convenient if we first express the number as a multiple of a power of 10, as follows:

$$25.00 = 2.500 \times 10^1$$

By **Equation 2.5** we then have

$$\log 25.00 = \log (2.500 \times 10^1) = \log 2.500 + \log 10^1$$

$$= 0.3979 + 1 = 1.3979$$

(*Note:* since log 1 = 0 and log 10 = 1—see Table 2.2—it is clear that log 2.500 must have a value between 0 and 1. The decimal point can therefore be fixed and log 2.500 = 0.3979.)

(b)
$$\log (2.500 \times 10^8) = \log 2.500 + \log 10^8$$

$$= 0.3979 + 8$$

$$= 8.3979$$

(c)
$$\log 0.02500 = \log (2.500 \times 10^{-2})$$

$$= \log 2.500 + \log 10^{-2}$$

$$= 0.3979 + (-2) = -1.6021$$

Example 2.6

Find the antilogarithm of (a) 1.3979 (b) 9.3979 (c) −1.6021.

Solution

(a) By **Equation 2.6**

$$\text{antilog } 1.3979 = \text{antilog } 1 \times \text{antilog } 0.3979$$

$$= 10 \times 2.500 = 25.00$$

(*Note:* since antilog of 0 is 1 and antilog of 1 is 10—see Table 2.2—it is clear that antilog of 0.3979 must be between 1 and 10. The decimal point can therefore be fixed; antilog 0.3979 = 2.500.)

(b)
$$\text{antilog } 8.3979 = \text{antilog } 8 \times \text{antilog } 0.3979$$

$$= 10^8 \times 2.500 = 2.500 \times 10^8$$

(c)
$$\text{antilog } (-1.6021) = ?$$

The first step in finding the antilog of a negative number is to rearrange it so that the part of the number after the decimal point has a positive value. −1.6021 can be written as −2 + 0.3979 (−2 + 0.3979 is normally written as $\bar{2}.3979$).

∴
$$\text{antilog } (-1.6021) = \text{antilog } (-2) \times \text{antilog } (0.3979)$$

$$= 10^{-2} \times 2.500 = 2.500 \times 10^{-2}$$

Example 2.7

The pH of a solution is defined by the equation

$$pH = -\log_{10} [H^+]$$

where $[H^+]$ denotes the hydrogen ion concentration (given in the units mol dm^{-3}). Calculate the pH of a solution whose hydrogen ion concentration (in mol dm^{-3}) is (a) 1.00 (b) 10.00 (c) 0.0100 (d) 1.26×10^{-4}.

Solution

We start with the defining equation for the required quantity. The defining equation shows that to calculate pH all we need to know is the hydrogen ion concentration $[H^+]$. This is given in the data. The pH can therefore be calculated, in each case, by the direct application of the defining equation.

(a) By definition, $pH = -\log_{10} [H^+]$
From data $[H^+] = 1.00$

$$\therefore \qquad pH = -\log_{10} 1$$
$$= 0$$

(b) $pH = -\log_{10} 10 = -1$
(c) $pH = -\log_{10} 0.01 = -\log_{10} 10^{-2} = -(-2) = 2$
(d) $pH = -\log_{10} (1.26 \times 10^{-10}) = -(\log 1.26 + \log 10^{-4})$
$$= -(0.1004 - 4) = 3.8996$$

Note

At first sight, this problem may have appeared difficult. This difficulty, however, is not real; it arises merely because *one often confuses unfamiliarity with difficulty.* Even if the concept of pH is not familiar, it does not matter as far as the calculation is concerned. To calculate pH all that is needed (see definition of pH) is the ability to find logarithms.

As a general rule, don't be put off by an unfamiliar problem. Instead, concentrate on analysing the problem logically, and on applying the relevant fundamental principles.

Example 2.8

Calculate the hydrogen ion concentration of a solution which has a pH value (a) 4.76 (b) −0.84.

Solution

We have to calculate $[H^+]$ from pH data. To do so we must start with an equation connecting $[H^+]$ and pH. The required equation was given in Example 2.7 and is $pH = -\log [H^+]$. By the direct application of this equation we can calculate $[H^+]$ since pH is given. The units for $[H^+]$ will be mol dm^{-3} by the definition of pH (see Example 2.7).

(a) From data pH = 4.76

$$\therefore \qquad 4.76 = -\log_{10}[H^+]$$
$$\therefore \log_{10}[H^+] = -4.76$$
$$\therefore \qquad [H^+] = \text{antilog}\ (-4.76)$$
$$= \text{antilog}\ (-5 + 0.24)$$
$$= \text{antilog}\ (-5) \times \text{antilog}\ 0.24 \qquad \text{(by Equation 2.6)}$$
$$= 10^{-5} \times 1.738 = 1.738 \times 10^{-5}\ \text{mol dm}^{-3}$$

(b) $pH = -\log[H^+]$

$$\therefore -0.84 = -\log[H^+]$$
$$\therefore \quad [H^+] = \text{antilog}\ 0.84$$
$$= 6.918\ \text{mol dm}^{-3}$$

Exercise 2.8

The pH of a solution is defined by the equation $pH = -\log_{10}[H^+]$. Calculate the $[H^+]$ of a solution whose pH is (a) 3.00 (b) −1.00 (c) 4.74 (d) 14.5.

Exercise 2.9

The pH of a solution ($pH = -\log_{10}[H^+]$) is 4.74. Calculate the electrode potential E of a hydrogen electrode in this solution given that $E = 0.0256\ \text{V} \times \log_e[H^+]$.

2.4 SI units

A physical quantity, in general, has both a number and a unit. We may therefore write

$$\text{physical quantity} = \text{number} \times \text{unit}$$

A number alone (i.e. without unit) for a physical quantity has no meaning (unless, of course, the physical quantity considered has no unit—this happens when the physical quantity is a ratio). For example, it would be meaningless to state that the length of an object is 2 without specifying the units (whether metres, feet, inches, etc.) in which length had been measured. The units for a physical quantity must therefore always be specified.

By international agreement a system of units known as the SI units (SI is an

Table 2.3   SI units of basic physical quantities

Name of quantity	Symbol for quantity	Name of SI unit	Symbol for SI unit
length	l	metre	m
mass	m	kilogram	kg
time	t	second	s
electric current	I	ampere	A
temperature	T	kelvin	K
amount of substance	n	mole	mol
luminous intensity	I_v	candela	cd

abbreviation for System Internationale) has now been recommended for all scientific work. The SI units are systematic, coherent, and are easy to use.

Table 2.3 gives the SI units, and their symbols, of the *basic physical quantities*. Study and memorize the data given in the table.

Note particularly that the SI unit for *amount of substance* is the *mole*. One mole of a substance is the amount of substance that contains 6.022×10^{23} particles; the nature of the particles considered must always be specified. For example, a mole of water contains 6.022×10^{23} water molecules, a mole of electrons contains 6.022×10^{23} electrons, and a mole of Cu^{2+} ions contains 6.022×10^{23} Cu^{2+} ions.

The SI unit for *temperature* is Kelvin (symbol K). Kelvin temperature is the 'absolute' temperature:

$$0\,°C = 273.15\,K \quad \text{and} \quad 100\,°C = 373.15\,K$$

Table 2.3 gives the basic physical quantities and their units. All other physical quantities can be derived from the basic physical quantities—they are therefore known as *derived quantities*. Some of the common derived physical quantities and their SI units are given in Table 2.4.

Read through Table 2.4 carefully. Remember that the units of a physical quantity follow directly from the definition of that quantity; how the units are derived from the definition will now be illustrated with some examples.

Table 2.4   Some derived quantities and their SI units

Name of quantity	Symbol for quantity	SI units (from definition)	Special name for SI unit	Other common units	Conversion factor
volume	V	m^3		litre (l)	$1\,l = 10^{-3}\,m^3$
density	ρ	$kg\,m^{-3}$		$g\,cm^{-3}$	$1\,g\,cm^{-3} = 10^3\,kg\,m^{-3}$
concentration	c	$mol\,m^{-3}$		$mol\,dm^{-3}$	$1\,mol\,dm^{-3} = 10^3\,mol\,m^{-3}$
molal concentration	m	$mol\,kg^{-1}$			
force	F	$kg\,m\,s^{-2}$	newton (N)	dyne	$1\,dyne = 10^{-5}\,N$
pressure	P	$N\,m^{-2}$ or $kg\,m^{-1}\,s^{-2}$	pascal (Pa)	atm Torr	$1\,atm = 1.013 \times 10^5\,Pa$ $1\,Torr = 1.33 \times 10^2\,Pa$
energy	E	$N\,m$	joule (J)	erg cal	$1\,erg = 10^{-7}\,J$ $1\,cal = 4.186\,J$
power	W	$J\,s^{-1}$	watt (W)		
electric potential	V	$W\,A^{-1}$	volt (V)		
electric charge	Q	$A\,s$	coulomb (C)		
resistance	R	$V\,A^{-1}$	ohm (Ω)		

Concentration (*c*) is defined as the amount of solute (moles) per unit volume of solution. That is

$$c = \text{amount/volume}$$

The SI unit for volume is metre³ (m^3) and therefore the SI unit for concentration *c* is

$$c = \frac{\text{mol}}{m^3} = \text{mol m}^{-3}$$

The most common unit of volume, however, is the dm^3 and concentration is usually expressed as mol dm^{-3}. Remember that units can always be subjected, like a number, to all the arithmetical and algebraic operations, and therefore

$$\frac{1}{m^3} = m^{-3} \left(\text{compare } \frac{1}{10^3} = 10^{-3} \right)$$

Force (*F*) is defined as mass × acceleration. Acceleration is the rate of change of velocity. The SI unit for velocity is metre/second = m/s = m s⁻¹. The SI unit for acceleration (acceleration = change in velocity/time) is therefore m s⁻¹/s = m s⁻². The SI unit for force (*F*) is therefore

$$F = \text{mass} \times \text{acceleration}$$

$$= \text{kg m s}^{-2}$$

The SI unit for force, kg m s⁻², is given a specific name *newton* (symbol N). That is, 1 N = −1 kg m s⁻².

Pressure (*p*) is defined as the force per unit area. That is

$$p = \text{force/area}$$

The SI unit for pressure is therefore

$$p = \frac{\text{newton}}{\text{metre}^2}$$

$$= \frac{N}{m^2} = \text{N m}^{-2} = \text{kg m}^{-1} \text{s}^{-2}$$

Often it is necessary to convert a physical quantity from one unit to another. To do so it is necessary to know the *conversion factor*, i.e. the quantitative relationship between the two units. Some conversion factors are given in the last column of Table 2.4.

In the SI system, various prefixes have been recommended for decimal multiples or submultiples of a unit. These prefixes are particularly useful when the physical quantity has either large or small numerical values in SI units. Some of the common prefixes are given in Table 2.5.

In a calculation it is always advisable to insert the units of a physical quantity along with the numerical value at each step in the calculation. This would help us to

Table 2.5   Recommended prefixes

Multiplication factor	Prefix	Symbol		Example
10^6	mega	M	megohm (MΩ)	1 MΩ = 10^6 Ω
10^3	kilo	k	kilometre (km)	1 km = 10^3 m
10^{-1}	deci	d	decimetre (dm)	1 dm = 10^{-1} m
10^{-2}	centi	c	centimetre (cm)	1 cm = 10^{-2} m
10^{-3}	milli	m	milligram (mg)	1 mg = 10^{-3} g
10^{-6}	micro	μ	microsecond (μs)	1 μs = 10^{-6} s
10^{-9}	nano	n	nanometre (nm)	1 nm = 10^{-9} m

avoid the mistake of being inconsistent in the choice of units. The units of the answer could also then be obtained by subjecting the units to the normal arithmetic and algebraic operations. Furthermore, if the units of the answer are incorrect, it follows that the method of calculation is incorrect. *You should cultivate the habit of inserting the units along with the numerical value at every step in a calculation.* This has been done in the worked examples in this book.

Example 2.9

The concentration (c) of a solute in a solution is 1.2×10^{-1} mol dm^{-3}. Express this concentration in SI units (1 dm = 10^{-1} m).

Solution
The problem is to find c in SI units. To be able to do so we should, first of all, know the SI unit for concentration. The SI unit is mol m^{-3}.

The problem, therefore, is to convert concentration given in mol dm^{-3} to mol m^{-3}. A look at the two units indicates that to convert mol dm^{-3} to mol m^{-3} all that is required is the relationship between dm^{-3} and m^{-3}. So let us first derive this relationship. It is given that

$$1 \text{ dm} = 10^{-1} \text{ m}$$

$$\therefore \quad 1 \text{ dm}^3 = (10^{-1} \text{ m})^3 = 10^{-3} \text{ m}^3$$

$$\therefore \quad 1 \text{ dm}^{-3} = \frac{1}{\text{dm}^3} = \frac{1}{(10^{-3} \text{ m}^3)}$$

$$= 10^3 \text{ m}^{-3} \qquad (1)$$

Equation (1), which shows the relationship between dm^{-3} and m^{-3}, can be made use of to convert mol dm^{-3} to mol m^{-3}. Consider the concentration (c) given in the data

$$c = 1.2 \times 10^{-1} \text{ mol dm}^{-3} \qquad (2)$$

On substituting for dm^{-3} in equation (2) by equation (1) we obtain

$$c = 1.2 \times 10^{-1} \text{ mol } (10^3 \text{ m}^{-3})$$

$$= 1.2 \times 10^2 \text{ mol m}^{-3}$$

which is the required answer.

Note

Understand clearly and remember the logic of the method. The problem is to eliminate dm^{-3} from equation (2) and substitute in its place the required unit m^{-3}. To do so, all that we need to know is the quantitative relationship between dm^{-3} and m^{-3} (this is given by equation 1).

Example 2.10

When one mole of hydrochloric acid, HCl (aq), is completely neutralized by a base the enthalpy change ΔH is given by $\Delta H = -13.7$ kcal mol^{-1}. Express this enthalpy change in SI units (1 kcal $= 4.18 \times 10^3$ J).

Solution

The SI unit for energy (see Table 2.3) is the joule (symbol J). Our problem therefore is to convert kcal into J.

$$\Delta H = -13.7 \text{ kcal mol}^{-1} \tag{1}$$

$$1 \text{ kcal} = 4.18 \times 10^3 \text{ J} \tag{2}$$

On substituting for kcal in equation (1) by equation (2) we get

$$\Delta H = -13.7 \, (4.18 \times 10^3 \text{ J}) \text{ mol}^{-1}$$

$$= -5.73 \times 10^4 \text{ J mol}^{-1}$$

Example 2.11

The earth's atmosphere exerts a pressure. The magnitude of this pressure depends on the altitude and varies from place to place. At sea level the pressure is approximately equal to that exerted by a column of mercury 0.760 metres high.

One atmosphere pressure (atm) is *defined* as the pressure exerted by a column of mercury 0.7600 metres high. Express 1 atm in SI units (density of mercury $= 1.359 \times 10^4$ kg m^{-3}, acceleration due to gravity $= 9.807$ m s^{-2}).

Solution

By definition

$$1 \text{ atm} = \text{the pressure exerted by a column of Hg, 0.7600 m high}$$

This definition shows that to convert 1 atm into SI units we must calculate the pressure, in SI units, exerted by a column of Hg 0.7600 m high.

How does the pressure exerted by a liquid column depend on its height? To answer this question we have to derive an equation which relates pressure due to a liquid column to its height. To do so we start with the defining equation for pressure and go on simplifying stepwise until we obtain an equation which shows the required relationship between pressure and height.

Pressure p is defined as the force per unit area. That is

$$p = \frac{\text{force}}{\text{area}}$$

$$= \frac{\text{mass} \times \text{acceleration}}{\text{area}} \quad (\text{since force} = \text{mass} \times \text{acceleration})$$

$$= \frac{\text{volume} \times \text{density} \times \text{acceleration}}{\text{area}} \left(\text{since density} = \frac{\text{mass}}{\text{volume}} \right)$$

On substituting for volume by the equation volume = height × area, and on simplification we get

$$p = \text{height} \times \text{density} \times \text{acceleration} \qquad (1)$$

Equation (1) can be used to calculate the pressure p exerted by a liquid column of known height. The pressure exerted by a column of mercury 0.7600 metres high, which is defined as equal to 1 atm, can therefore be calculated by application of this equation. On substituting values for each physical quantity in SI units we get

$$1 \text{ atm} = (0.7600 \text{ m}) (1.359 \times 10^4 \text{ kg m}^{-3}) (9.807 \text{ m s}^{-2})$$

$$= 1.013 \times 10^5 \text{ kg m}^{-1} \text{s}^{-2} \quad \text{or} \quad 1.013 \times 10^5 \text{ Pa}$$

The above equation gives the required relationship between atm and the SI unit ($\text{kg m}^{-1} \text{s}^{-2}$, Pascal) for pressure.

Exercise 2.10

The concentration of a solute in a solution is 1.3×10^{-5} mol cm^{-3}. Express this concentration in (a) mol dm^{-3} (b) SI units.

Exercise 2.11

Density is defined as the mass per unit volume. The density of a gas is 0.080 g cm^{-3}. What is the density of the gas in SI units?

Exercise 2.12

The normal body temperature of a human adult is 98.4 °F. Give this temperature in (a) °C (b) SI units, given that 0 °C = 32 °F = 273 K and that 100 °C = 212 °F = 373 K.

3 | Atoms and molecules

3.1 Basic concepts

Relative atomic mass and relative molecular mass

Atoms and molecules are minute, and their masses are very small. Their masses are therefore generally given relative to one another. The carbon atom ^{12}C is now used as the standard and its relative mass is taken, by convention, to be exactly 12. The masses of atoms and molecules relative to this standard are known as relative atomic masses and relative molecular masses respectively.

Relative atomic mass (symbol, A_r) of an element is defined as the ratio of the mass of one atom of that element to 1/12th the mass of an atom of the ^{12}C isotope of carbon. That is

$$A_r = \frac{\text{mass of one atom}}{(1/12) \times \text{the mass of a } ^{12}C \text{ atom}}$$
Equation 3.1a

Relative molecular mass (symbol, M_r) is defined similarly by the ratio

$$M_r = \frac{\text{mass of one molecule}}{(1/12) \times \text{the mass of a } ^{12}C \text{ atom}}$$
Equation 3.1b

The terms relative atomic mass and relative molecular mass replace, in the SI system, the old terms atomic weight and molecular weight.

Mole, molar properties

Many physical properties (e.g. volume of a gas, osmotic pressure of a solution) depend on the number of particles (molecules, atoms, ions) in the system but not on the nature of the particles. It is advantageous, therefore, to express *the amount of a substance* in terms of the number of particles (the advantage of doing so will be seen more clearly after you solve Exercise 5.2).

6.022 $\times$ 10^{23} particles have been taken as the measure for the amount of substance, and this is called the mole (symbol mol). Why the number 6.022 $\times$ 10^{23}? This number was chosen because it so happens that 12.000 grams of the ^{12}C isotope is found by experiment to contain 6.022 $\times$ 10^{23} atoms. The formal definition of the mole is that it is the amount of substance of a system which contains as many elementary entities as there are atoms in 0.012 kg of ^{12}C.

This definition can be applied to any system of particles but the exact nature of the particles must be specified. Thus a mole of hydrogen molecules contains 6.022 $\times$ 10^{23}

H_2 molecules, a mole of electrons contains 6.022×10^{23} electrons, and a mole of SO_4^{2-} ions contains 6.022×10^{23} SO_4^{2-} ions.

The number (N) of particles per mole is known as the Avogadro constant (symbol, L). This statement can be given (see section 2.2) in equation form as

$$L = \frac{N}{n} \qquad \textbf{Equation 3.2}$$

where n is the moles corresponding to N particles. The value of the Avogadro constant is 6.022×10^{23} mol^{-1}.

Any property per mole is known as a molar property. For example, the mass per mole (i.e. the mass of one mole) is known as the *molar mass*, the volume per mole is the *molar volume*, the charge per mole is the *molar charge*, the energy per mole is the *molar energy*, and so on.

Since a molar property is the value of that property per mole we may write (see section 2.2)

$$\text{molar property} = \frac{\text{value of property}}{\text{number of moles}}$$

For example, since the molar mass (M) is the mass (m) per mole we may write

$$M = \frac{m}{n} \qquad \textbf{Equation 3.3}$$

Exercise 3.1

How many atoms are present in 1.200×10^{-2} g of the ^{12}C isotope of carbon?

Exercise 3.2

Derive an equation that can be used for calculating the number N of molecules present in a known mass m of a compound whose molar mass is known.

Indicate how you would calculate the mass of a molecule of the compound by making use of this equation.

Example 3.1 Determination of the Avogadro constant

One experimental method for the determination of the Avogadro constant is by a study of the decay of certain radioactive substances. Radium, for example, spontaneously disintegrates emitting high energy helium ions (He^{2+}). The number of these ions emitted may be counted by allowing them to bombard a screen coated with zinc sulphide. A greenish glow would then be seen (after magnification, for example by a telescope) each time a helium ion hits the screen.

Experiments show that 1.16×10^{18} helium particles are emitted from 1.00 g of radium in one year; the mass of helium gas then formed is 0.0077 mg. Calculate the Avogadro constant.

Solution

Step 1 Clarify and define the problem

We have to calculate the Avogadro constant L from the given data. The data give the mass m and the number N of helium particles emitted from a known mass of radium in a known time.

Step 2 Select the key equation

We start, as recommended in Chapter 1, with the definition of the required quantity. The Avogadro constant is the number of particles per mole. This definition could be given in equation form as

$$L = \frac{N}{n}$$

(1) **Equation 3.2**

Step 3 Derive the equation for the calculation

Equation (1) shows that to calculate L we require the number N of particles present in a known amount n of a substance.

Now look at the data to see whether these two quantities are given. N is given ($N = 1.16 \times 10^{18}$ helium particles); n which is the number of moles corresponding to N particles is not given. n is however related to the data given—the mass m of 1.16×10^{18} particles—by the equation

$$n = \frac{m}{M}$$

(2) **Equation 3.3**

where M is the molar mass of helium.

On substituting for n in equation (1) by equation (2) we get

$$L = \frac{NM}{m}$$

(3)

Equation (3) could be used for calculating L since all the quantities required for its calculation are known.

Step 4 Collect the data, check the units, and calculate

We collect and set out the data necessary for the calculation of L by equation (3)

$$N = 1.16 \times 10^{18}$$

$$m = 0.0077 \text{ mg}$$

$$M = 4.00 \text{ g mol}^{-1}$$

We now check the units. Since M is given in terms of grams (see note 1), it is convenient to express m also in grams (remember $1 \text{ mg} = 10^{-3}$ g)

$$m = 0.0077 \text{ mg}$$

$$= 0.0077 \times 10^{-3} \text{ g} \qquad \text{(see section 2.1)}$$

$$= 7.7 \times 10^{-6} \text{ g}$$

On calculation by equation (3) we obtain

$$L = \frac{(1.16 \times 10^{18})\,(4.00\text{ g mol}^{-1})}{(7.7 \times 10^{-6}\text{ g})}$$

$$= 6.02 \times 10^{23}\text{ mol}^{-1}$$

The Avogadro constant is found by this calculation to be $6.02 \times 10^{23}\text{ mol}^{-1}$.

Step 5 Review and check the solution

As a general rule we should, as emphasized in Chapter 1, quickly glance through and review all aspects of the problem and the solution. This is particularly to make sure that the answer obtained is for the problem posed, that the answer has the correct order of magnitude and is not unreasonable, that the units are correct, and that the number of significant figures in the answer is appropriate.

In this example, the answer obtained is for the problem posed and is correct (see Appendix 1), the units are correct, and the number of significant figures is appropriate.

Notes

1 It is shown in Example 3.3 that the molar mass M of the atom of an element is given by

$$M = A_r \times \text{g mol}^{-1}$$

where A_r is the relative atomic mass of the element. A_r values should be obtained, whenever required, from a table (see Appendix 2); the value for helium is 4.00. The molar mass of helium is therefore 4.00 g mol^{-1}.

2 Two items from the given data (mass of radium and time) are not required for the calculation. These unnecessary items of data do not cause confusion if the problem is analysed logically.

Example 3.2 Calculation of mass and size of atoms

By making use of the value for the Avogadro constant calculate (a) the mass (b) the approximate volume, of a silver atom. Molar mass of silver $= 107.87\text{ g mol}^{-1}$, density of silver $= 10.50\text{ g cm}^{-3}$.

Solution to Part (a)

Step 1 Clarify and define the problem

We have to calculate the mass of an individual silver atom, m_{Ag}, from the molar mass M and the Avogadro constant L. To do so we have to derive an equation that shows the relationship between m_{Ag} and M and L.

Step 2 Select the key equation

By the principle of *additivity of mass* it is clear that m_{Ag} is related to M by the equation

$$m_{Ag} = \frac{M}{\text{number of Ag atoms in one mole}} \tag{1}$$

Step 3 Derive the equation for the calculation

The denominator in equation (1) is, by definition, equal to L. Equation (1) can be rewritten

$$m_{Ag} = \frac{M}{L} \tag{2}$$

Step 4 Collect the data, check the units, and calculate

$$M = 107.87 \text{ g mol}^{-1}$$

$$L = 6.022 \times 10^{23} \text{ mol}^{-1}$$

On inserting these values in equation (2) and on calculating we obtain

$$m_{Ag} = \frac{107.87 \text{ g mol}^{-1}}{6.022 \times 10^{23} \text{ mol}^{-1}}$$

$$\therefore \qquad m_{Ag} = 1.791 \times 10^{-22} \text{ g}$$

Step 5 Review and check the solution

The answer obtained, the units, and the number of significant figures are all reasonable.

Solution to Part (b)

The volume of an individual atom is given by the following equation which is analogous to equation 2 (see note 2)

$$\text{volume of a Ag atom} = \frac{\text{volume of one mole of Ag}}{6.022 \times 10^{23}}$$

$$= \frac{\text{mass of one mole of Ag}}{\text{density} \times 6.022 \times 10^{23}} \left(\text{since volume} = \frac{\text{mass}}{\text{density}} \right)$$

$$= \frac{107.87 \text{ g mol}^{-1}}{(10.50 \text{ g cm}^{-3}) (6.022 \times 10^{23} \text{ mol}^{-1})}$$

$$= 1.70 \times 10^{-23} \text{ cm}^3$$

Notes

1 In solids and liquids the particles present (atoms, ions, or molecules) touch one another. The volume of a liquid or solid may therefore be assumed to be approximately equal to the sum of the volumes of the individual particles.
2 Because part (b) is so similar to part (a) we have not set out the five-step procedure in detail in the solution to part (b).

Example 3.3 Relationship between molar mass and relative atomic mass

Find the molar mass of oxygen atoms given that the relative atomic mass of oxygen $A_r(O)$, is 16.00. The molar mass of ^{12}C is 12.00 g mol^{-1}.

Solution

Step 1 *Clarify and define the problem*

We want the molar mass of oxygen atoms—let this be represented by M.

Step 2 *Select the key equation*

As key equation we select the defining equation for M which is

$$M = \frac{m}{n}$$ (1) **Equation 3.3**

Step 3 *Derive the equation for the calculation*

Analyse equation (1). It is seen that when $n = 1$ the mass m is equal to the molar mass M. That is

$$M = \text{mass of 1 mole of O atoms}$$

Since one mole of oxygen atoms contains 6.022×10^{23} O atoms we may rewrite the last equation as

$$M = \text{mass of } 6.022 \times 10^{23} \text{ atoms of O}$$

$$= \text{mass of 1 atom of O} \times 6.022 \times 10^{23}$$ (2)

The mass of one atom of O, required for the calculation of M by equation (2), is related to the data given ($A_r(O)$ of oxygen) by the equation

$$\text{mass of an O atom} = A_r(O) \times \frac{\text{mass of an atom of } ^{12}C}{12}$$ (3) **Equation 3.1**

Substitution of equation (3) in equation (2) then gives

$$M = \frac{A_r(O)}{12} \times \text{mass of 1 atom of } ^{12}C \times 6.022 \times 10^{23}$$

$$= \frac{A_r(O)}{12} \times \text{molar mass of } ^{12}C$$ (4)

Step 4 *Collect the data, check the units, and calculate*

From the given data

$$A_r(O) = 16.00$$

$$\text{molar mass of } ^{12}C = 12.00 \text{ g mol}^{-1}$$

Insertion of these values in equation (4) and calculation then shows that

$$M = \frac{16.00}{12} \times 12.00 \text{ g mol}^{-1}$$

$$= 16.00 \text{ g mol}^{-1}$$

Step 5 Review and check the solution

The solution is somewhat easy but the result obtained is important. Remember, as a general rule, that for any atom

$$\text{molar mass} = A_r \times \text{g mol}^{-1}$$

Similarly it is easily shown that for any molecule

$$\text{molar mass} = M_r \times \text{g mol}^{-1}$$

Exercise 3.3

How many (a) moles of H atoms (b) atoms of H are present in one mole of $CuSO_4 \cdot 5H_2O$?

Exercise 3.4

The mass of one atom of hydrogen is 1.67×10^{-24} g. Calculate the molar mass of (a) hydrogen atoms (b) hydrogen molecules.

Exercise 3.5

How many (a) moles (b) molecules are present in 1.00 cm³ of a liquid whose density is 0.750 g cm⁻³? The relative molecular mass of the liquid is 60.0.

Exercise 3.6

The molar volume of water is 18.0 cm³ mol⁻¹. Estimate the volume of an individual water molecule. Explain why you would expect this calculation to be approximate.

Exercise 3.7

Suppose that one mole of a compound A is dissolved in the ocean and mixed uniformly with all the water in it. Show that a 1 cm³ sample of the ocean water would then contain about 600 molecules of A. Assume that the total volume of water in the ocean is 10^{15} m³. (This exercise should help you to appreciate the enormous magnitude of the Avogadro Constant—6.022×10^{23} mol⁻¹.)

3.2 Chemical formulae

Two types of chemical formulae are empirical formulae and molecular formulae. An *empirical formula* indicates the simplest whole number ratio of the atoms present in a compound. A *molecular formula* indicates the number of each type of atom that is present in a molecule.

Chemical formulae are normally determined from percentage by mass data of the elements found by analysis of the compound. The method of deducing formulae from percentage by weight data is illustrated in Example 3.4.

The chemical formula of a compound contains useful quantitative information about that compound. The types of information that can be deduced from a molecular formula are illustrated in Example 3.5 for a simple molecule.

Example 3.4 Deduction of molecular formula from analytical data

Analysis of an organic compound X showed that it contained 40.0% by weight of carbon, 6.7% by weight of hydrogen, and 53.3% by weight of oxygen. The molar mass of X is 180 g mol^{-1}. Deduce the molecular formula of X.

Solution

Step 1 Clarify and define the problem

The compound X contains C, H, and O only; to find its molecular formula we should determine the number of atoms of each of the elements C, H, and O present in one molecule of X.

Since the number of atoms N is directly proportional to the moles n (see **Equation 3.2**, $N = n \times 6.022 \times 10^{23}$), it is clear that we could also find the molecular formula by determining the moles of each element present in one mole of the compound.

Step 2 Select the key equation

To calculate the moles of an element n_e present in one mole of the compound we can make use of the equation

$$n_e = \frac{m_e}{M_e} \qquad \text{(1) Equation 3.3}$$

Here m_e would denote the mass of the element present in one mole of the compound, and M_e is the molar mass of the element.

Step 3 Derive the equation for the calculation

To calculate n_e by equation (1) we require m_e—the mass of the element present in one mole of the compound. This could be done by making use of the percentage by weight (% wt) data given. The percentage by weight (% wt) of an element in a compound is defined by the equation:

$$\% \text{ wt} = \frac{\text{mass of element in the compound}}{\text{mass of compound}} \times 100 \qquad \textbf{Equation 2.3}$$

By rearrangement of Equation 2.3 the mass of element m_e present in one mole of the compound X is given by

$$m_e = \frac{\% \text{ wt} \times \text{mass of one mole of compound}}{100} \qquad (2)$$

The mass of one mole of the compound, m_c, is related to the molar mass of the compound M_c by **Equation 3.3**. On replacing m in equation (1) by equation (2) we obtain

$$n_e = \frac{\% \text{ wt} \times m_c}{M_e \times 100} \qquad (3)$$

Equation (3) can be used to calculate the moles m_e of each of the elements present in one mole of the compound.

Step 4 Collect the data, check the units, and calculate

It is given that $M_c = 180$ g mol^{-1} and so $m_c = 180$ g (**Equation 3.3**).

$$\% \text{ wt of carbon} = 40.0$$

$$\% \text{ wt of hydrogen} = 6.7$$

$$\% \text{ wt of oxygen} = 53.3$$

The relative atomic masses of the elements C, H, and O present in the compound X can be looked up from a table (see Appendix 2) and are 12.01, 1.008, and 16.00 respectively. The molar masses of C, H, and O are therefore (see Example 3.3)

$$M_{\text{carbon}} = 12.01 \text{ g mol}^{-1}$$

$$M_{\text{hydrogen}} = 1.008 \text{ g mol}^{-1}$$

$$M_{\text{oxygen}} = 16.00 \text{ g mol}^{-1}$$

By equation (3) the moles of carbon n_C present in one mole of the compound X is

$$n_C = \frac{40.0 \times 180 \text{ g}}{12.01 \text{ g mol}^{-1} \times 100}$$

$$= 6 \text{ mol}$$

Similarly it is easily seen that

$$n_H = 12 \text{ mol} \quad \text{and} \quad n_O = 6 \text{ mol}$$

That is, one mole of X contains 6 moles of C, 12 moles of H, and 6 moles of O.

∴   molecular formula of X is $C_6H_{12}O_6$

Step 5 Review and check the solution

Note

If the molar mass of the compound is not known, it is not possible to obtain the molecular formula. Only the empirical formula can then be obtained.

In the above example, if the molar mass of the compound X is not given, we can find only the empirical formula. The calculation is similar to that given in Example 3.4; the steps are tabulated below.

	C	:	H	:	O
weight ratio (from data)	40.0	:	6.7		53.3
∴ molar ratio	$\frac{40.0}{12.01}$	:	$\frac{6.7}{1.008}$	:	$\frac{53.3}{16.00}$
(by **Equation 3.3**)					
=	3.33	:	6.65	:	3.33
∴ simplest whole number molar ratio	1	:	2	:	1
∴ simplest whole number ratio of the atoms present	1	:	2	:	1

The empirical formula, therefore, is $C_1H_2O_1 = CH_2O$

Example 3.5 Calculation based on molecular formula

Calculate the percentage by weight of carbon, hydrogen, and oxygen in glucose which has the formula $C_6H_{12}O_6$. The relative atomic masses (A_r) of C, H, and O are 12.01, 1.008, and 16.00 respectively.

Solution

Step 1 Clarify and define the problem

We have to calculate the percentage by weight of the elements present in a compound.

Step 2 Select the key equation

We start, as usual, with the defining equation for the required quantity. The percentage by weight of an element in a compound (% wt) is defined by the equation (see **Equation 2.3**)

$$\% \text{ wt} = \frac{m_e}{m_c} \times 100 \qquad \text{(1) Equation 2.3}$$

Step 3 Derive the equation for the calculation

To calculate % wt using equation (1) we should find the mass of the element (m_e) present in a known mass of the compound (m_c). We are free to consider any mass of the compound and find the mass of the element present in it.

Consider the molar mass of the compound (M_c).

$$m_c = M_c \times n_c \qquad \text{(2) Equation 3.3}$$

The mass of each element (m_e) present in the molar mass of the compound can be calculated by making use of the equation

$$m_e = M_e \times n_e \qquad \text{(3) Equation 3.3}$$

M_e is the molar mass of the element in the compound and n is the moles of the element present in one mole of the compound.

On replacing m_c and m_e in equation (1) by equations (2) and (3) we obtain

$$\% \text{ wt} = \frac{n_e \times M_e}{n_c \times M_c} \times 100 \qquad \text{(4)}$$

Equation (4) can be used to calculate the percentage by weight (% wt) of each of the elements present in a compound.

Step 4 Collect the data, check the units, and calculate

$$M_c = (6 \times 12.01 + 12 \times 1.008 + 6 \times 16.00) \text{ g mol}^{-1}$$

$$= 180.16 \text{ g mol}^{-1}$$

$$M_{\text{carbon}} = 12.01 \text{ g mol}^{-1}$$

$$n_{\text{carbon}} = 6 \text{ mol (from formula } C_6H_{12}O_6)$$

$$n_c = 1 \text{ mol}$$

$$\therefore \quad \% \text{ wt of carbon} = \frac{6 \text{ mol} \times 12.01 \text{ g mol}^{-1} \times 100}{1 \text{ mol} \times 180.16 \text{ g mol}^{-1}}$$

$$= 40.0$$

A similar application of equation (4) shows that

$$\% \text{ wt of hydrogen} = 6.7$$

$$\% \text{ wt of oxygen} = 53.3.$$

The percentage by weight of carbon, hydrogen, and oxygen in $C_6H_{12}O_6$ is therefore 40.0, 6.7, and 53.3 respectively.

Step 5 Review and check the solution

Example 3.5 is the reverse of Example 3.4 and the figures correspond.

Exercise 3.8

Ilmenite, a black beach sand, is an important mineral because it contains the useful metal titanium (Ti). The formula of ilmenite is $FeO \cdot TiO_2$. Calculate
  (a) the moles of titanium atoms in 100.0 g of ilmenite
  (b) the percentage by weight of titanium in ilmenite.

Exercise 3.9

Sri Lanka is well known for excellent gem stones such as sapphire, ruby, amethyst, cat's eye, topaz, moonstone, etc. Chemical analysis shows the following composition by weight

  (a) sapphire: aluminium (Al) 53%; oxygen (O) 47%
  (b) chrisoberyl (cat's eye): beryllium (Be) 7.1%; Al 42.5%; O 50.4%.

Find the simplest possible formula of sapphire and chrisoberyl.

Exercise 3.10

In the nineteenth century one method used (Cannizzaro's method) for determining the molar mass of an element was to determine, by chemical analysis, the composition of various compounds of that element. If a large number of compounds are analysed it is almost certain

Table 3.1   Data for Exercise 3.10

Compound	% by weight of oxygen	Molar mass of compounds/g mol⁻¹
water	88.1	18
carbon dioxide	72.8	44
dinitrogen oxide (N₂O)	36.4	44
nitrogen monoxide (NO)	53.3	30

that one of them (at least) will contain only one atom of the element in the molecule. In such a case the mass of the element present in the molar mass of the compound would evidently be the molar mass of the element.

Use the data given in Table 3.1 to determine the molar mass of oxygen atoms.

Exercise 3.11

Chlorophyll, the important green pigment in plants which catalyses the manufacture of sugars from carbon dioxide, water, and sunlight has 2.68% by weight of magnesium. Each chlorophyll molecule has one atom of magnesium. Calculate the molar mass of chlorophyll.

Exercise 3.12

1.500 g of an organic compound containing C, H, and O only is burnt in excess oxygen to form 3.300 g of CO_2 and 1.800 g of H_2O. The molar mass of the compound is 60.0 g mol^{-1}. Find the molecular formula of the compound.

Exercise 3.13

The average percentage composition, by *volume*, of dry air is:

N_2 78.09; O_2 20.95; Ar 0.93; CO_2 0.030; Ne 1.8×10^{-3};

He 5.2×10^{-4}; Kr 1×10^{-4}; H_2 5×10^{-5}; O_3 5×10^{-5}; Xe 3×10^{-6}

By assuming that the volume of a gas is directly proportional to the number of moles and is independent of the nature of the gas, calculate the percentage composition, by *mass*, of dry air.

Review of Chapter 3

Three fundamental equations were introduced in this chapter. You should understand, and also be able to recall and apply, these equations whenever necessary. The equations are

$$A_r \text{ (or } M_r) = \frac{\text{mass of 1 atom (or 1 molecule)}}{(1/12) \times \text{mass of a } {}^{12}C \text{ atom}} \qquad \textbf{3.1a, 3.1b}$$

$$L = \frac{N}{n} \qquad \textbf{3.2}$$

$$M = \frac{m}{n} \qquad \textbf{3.3}$$

where A_r is the relative atomic mass
M_r is the relative molecular mass
L is the Avogadro constant
N is the number of particles
n is the amount of substance
M is the molar mass
m is the mass

4 | Chemical equations

4.1 Information provided by a balanced equation

A balanced equation for a chemical reaction shows the ratio in which atoms, molecules, or ions take part in the reaction. For example, the balanced equation

$$2Mg(s) + O_2(g) \rightarrow 2MgO(s)$$

shows that two moles of magnesium atoms in the solid state react with one mole of oxygen molecules in the gaseous state to give two moles of magnesium oxide (MgO) in the solid state.

A balanced equation is very important because we can deduce from it a large amount of useful information. For example, the equation can be interpreted in terms of moles, masses, or, for gases, volumes.

Since the *moles* (n) of any substance is directly proportional to the number of particles N (see **Equation 3.2**, $N = n \times 6.022 \times 10^{23}$), it follows that in the balanced equation given, two moles of magnesium atoms react with one mole of oxygen molecules to give two moles of magnesium oxide.

The *masses* taking part in a reaction are also easily calculated: moles (n) can be converted into masses (m) by using the equation $m = n \times$ molar mass (see **Equation 3.3**).

The *volume* of an ideal gas involved in a reaction is also easily calculated. This calculation depends on the fact that volumes (V) can be calculated from moles (n) by using the ideal gas equation from which we get $V = nRT/p$ (see **Equation 5.1**).

Example 4.1 Calculations based on a balanced equation

Ammonia (NH_3) is a very important compound. It is used for the manufacture of nitrogenous fertilizers, nylon, and other important substances. The industrial manufacture of ammonia is by the Haber process which makes use of the almost inexhaustible supply of nitrogen in the atmosphere.

1.00 mol of nitrogen molecules and 1.50 mol of hydrogen molecules are mixed together in a closed vessel. If the reaction goes to completion, calculate the moles of each gas that would then be present in the final mixture.

Solution

Step 1 Clarify and define the problem

We have to find the composition of the mixture after a chemical reaction given the initial composition. The change in the moles of each reactant and product due to the reaction must therefore first be found. This can be done by making use of the balanced equation for the reaction.

Step 2 Select the key equation

The key equation for the calculation is the balanced equation for the reaction which is

$$N_2(g) + 3H_2(g) \rightarrow 2NH_3(g)$$

Step 3 Derive the equation for the calculation

The equation for the calculation is the balanced equation which shows that one mole of N_2 would react with three moles of H_2 to give two moles of NH_3.

Step 4 Collect the data, check the units, and calculate

The data given and the steps in the calculation are best tabulated as in Table 4.1 (see note 2).

Table 4.1   Data and calculations for Example 4.1

Balanced equation	N_2	$+$	$3H_2$	$\longrightarrow$	$2NH_3$	Comments
moles present initially	1.00		1.50		0	given in data
moles that react	0.50		1.50		—	deduced from the balanced equation
moles formed	—		—		1.00	deduced from the balanced equation
moles present after reaction	$(1.00 - 0.50)$ $= 0.50$		$(1.50 - 1.50)$ $= 0$		1.00	

The final mixture would therefore contain 0.50 mol N_2 and 1.00 mol NH_3.

Step 5 Review and check the solution

Notes

1 The balanced equation indicates that H_2 and N_2 would always react in the molar ratio 3:1. To react with the 1.50 moles of H_2 present initially only 0.50 mole of N_2 is therefore required. N_2 is thus present in excess; some unreacted N_2 would therefore be present in the final mixture.

2 The steps in the calculation have been tabulated. Tabulation helps in the presentation of the problem in a clear manner, and it also guides us through the calculation (read section 1.2). At a glance we can then 'see' the whole problem and the solution. In calculations based on a balanced equation tabulation is strongly recommended.

Example 4.2 Calculation of the molar mass by making use of a balanced equation

0.486 g of a metal X when burnt in excess oxygen gave 0.806 g of its oxide XO. Calculate the molar mass of X.

Solution

Step 1 Clarify and define the problem

The required quantity is the molar mass of X. Call it M_x.

Step 2 Select the key equation

M_x could be calculated by making use of the information provided by the balanced equation for the reaction which is

$$X + \tfrac{1}{2}O_2 \rightarrow XO \tag{1}$$

and by the equation for molar mass

$$M_x = \frac{m_x}{n_x} \tag{2} \textbf{ Equation 3.3}$$

Step 3 Derive the equation for the calculation

Equation (1) indicates that one mole of X reacts with $\frac{1}{2}$ mole of O_2 molecules or with one mole of oxygen atoms.

That is

$$n_x = n_O \tag{3}$$

n_O is the moles of oxygen atoms

$$n_O = \frac{M_O}{m_O} \tag{4} \textbf{ Equation 3.3}$$

Combining (2), (3), and (4) we have

$$\frac{M_x}{m_x} = \frac{M_O}{m_O}$$

$$M_x = \frac{M_O m_x}{m_O}$$

We know m_x and m_O and we can look up M_O from a table of relative atomic masses because $M = A_r \times$ g mol^{-1} (see note to Example 3.1, page 31).

Step 4 Collect the data, check the units, and calculate

From the given data it is seen that

$$m_x = 0.486 \text{ g}$$

$$m_O = (0.806 - 0.486) \text{ g} = 0.320 \text{ g}$$

$$M_O = 16.00 \text{ g mol}^{-1}$$

$$M_x = \frac{16.00 \text{ g mol}^{-1} \times 0.486 \text{ g}}{0.320 \text{ g}}$$

$$= 24.3 \text{ g mol}^{-1}$$

$\therefore$ molar mass of X $= 24.3$ g mol^{-1}

Step 5 Review and check the solution

Note

This example illustrates one method that was widely used in the nineteenth century for determining molar masses (then called atomic weights) of elements.

Exercise 4.1

Consider Example 4.2. How many (a) moles of oxygen molecules (b) moles of oxygen atoms (c) molecules of oxygen, reacted with 0.486 g of the metal?

Exercise 4.2

When 0.50 mol of nitrogen dioxide NO_2, is heated in a closed vessel it partly dissociates according to the equation $2NO_2 \rightarrow 2NO + O_2$. If 0·18 mol of nitrogen dioxide dissociates, calculate (a) the mols of NO_2, NO, and O_2 (b) the mass of each of these gases in the reaction vessel. Tabulate your results.

Exercise 4.3

The reactions in the industrial manufacture of sulphuric acid by the contact process are

$$S + O_2 \rightarrow SO_2$$

$$2SO_2 + O_2 \rightarrow 2SO_3$$

$$SO_3 + H_2O \rightarrow H_2SO_4$$

What mass of sulphur is necessary to produce 1.000 kg of H_2SO_4, assuming 100% conversion?

Exercise 4.4

When finely ground limestone ($CaCO_3$) and clay ($Al_2O_3 \cdot 2SiO_2 \cdot 2H_2O$) are roasted together at a high temperature (about 1450 °C) in a rotary kiln, cement clinker is produced. The more important chemical reactions then taking place are

$$CaCO_3 \rightarrow CaO + CO_2$$

$$Al_2O_3 \cdot 2SiO_2 \cdot 2H_2O \rightarrow Al_2O_3 + 2SiO_2 + 2H_2O$$

$$3CaO + SiO_2 \rightarrow 3CaO \cdot SiO_2$$

$$2CaO + SiO_2 \rightarrow 2CaO \cdot SiO_2$$

$$3CaO + Al_2O_3 \rightarrow 3CaO \cdot Al_2O_3$$

Calculate the mass of limestone necessary to convert all the silica in 25.0 g of clay into tricalcium silicate ($3CaO \cdot SiO_2$) and dicalcium silicate ($2CaO \cdot SiO_2$). Assume that 70% by weight of the SiO_2 is converted into tricalcium silicate.

4.2 Method of balancing equations

A balanced equation provides important information. You should therefore understand and remember the method of balancing equations.

Before an equation can be balanced the reactants and products must be known. These are experimental quantities. Once the reactants and products are known *balancing is always done by the application of the laws of conservation of matter and of charge.*

The *law of conservation of matter* states that matter cannot be created or destroyed. From this law it follows that the total number of atoms of each kind must be the same on either side of an equation for a chemical reaction.

The *law of conservation of charge* states that charge cannot be created or destroyed. According to this law the same net charge must be present on either side of any equation.

Example 4.3   Illustration of method of balancing equations

The main constituent of petrol (gasoline) is octane, C_8H_{18}. Its combustion in the engine of motor vehicles provides the energy necessary for motion. The products of combustion are water (H_2O) and carbon dioxide (CO_2). Write the balanced equation for the reaction.

Solution

The reactants and products are given in the data; the conversion of reactants into products may be written as

$$C_8H_{18} + O_2 \rightarrow CO_2 + H_2O$$

This equation contradicts the law of conservation of matter because the number of atoms of each kind is not the same on either side. The equation, therefore, is not balanced.

Our problem is to balance this equation. This could be done by balancing each kind of atom in a stepwise manner. Consider first the C atoms. On balancing them one obtains

$$C_8H_{18} + O_2 \rightarrow 8CO_2 + H_2O$$

The hydrogen and oxygen atoms have now to be balanced. On balancing H atoms we get

$$C_8H_{18} + O_2 \rightarrow 8CO_2 + 9H_2O$$

Only the O atoms are now not balanced. To balance them we must write

$$C_8H_{18} + 12.5O_2 \rightarrow 8CO_2 + 9H_2O \tag{1}$$

All coefficients in the above equation could be given as integers by multiplying the above equation by 2

$$2C_8H_{18} + 25O_2 \rightarrow 16CO_2 + 18H_2O \tag{2}$$

Equations (1) and (2) are balanced equations. It will be seen that the laws of conservation are satisfied; the same number of each kind of atom, and the same charge (zero charge) is present on either side of the equation.

Exercise 4.5

How many grams of oxygen are necessary for the complete combustion of 1.00 mol of molecules of ethane, C_2H_6?

4.3 Balancing acid–base reactions in aqueous solution

In all acid–base reactions occurring in aqueous solution the *only* chemical reaction taking place is

$$H^+(aq) + OH^-(aq) \rightarrow H_2O(l)$$

H^+ ions from the acid (all acids produce H^+ ions in aqueous solution) react with OH^- ions from the base (all soluble bases produce OH^- ions in aqueous solution) to give H_2O. The other ions produced from the acid or base do not take part in the reaction.

The balancing of acid–base reactions, like that for all reactions, is based on the application of the laws of conservation of mass and of charge. The problem, however, is simplified by first breaking it down into simpler problems. Any acid–base reaction, in aqueous solution, can be considered to be the sum of the following three simpler processes

(a) the production of H^+ ions from the acid
(b) the production of OH^- ions
(c) the reaction between H^+ and OH^- ions to form H_2O.

To write the balanced equation for an acid–base reaction the simplest procedure therefore is first to write the balanced equations for the three simpler processes and then add them together. This method is illustrated in Example 4.4.

Example 4.4 Writing a balanced equation for an acid–base reaction in aqueous solution

Write the balanced equation for the reaction, in aqueous solution, between (a) $KHC_2O_4 \cdot H_2C_2O_4$ and $NaOH$ (b) HCl and Na_2CO_3.

Solution
To write the balanced equation for any reaction we must first know the reactants and products. The products have not been given in the problem statement. This is because we must remember that in all acid–base reactions, in aqueous solution, the only chemical reaction taking place is

$$H^+(aq) + OH^-(aq) \rightarrow H_2O(l)$$

The procedure outlined in the text for writing the balanced equation for an acid–base reaction is applied below for writing the required balanced equations.
(a) The production of H^+ ions from the acid $KHC_2O_4 \cdot H_2C_2O_4$ may be written as:

$$K^+(aq) + 3H^+(aq) + 2C_2O_4^{2-}(aq) \tag{1}$$

It is seen that one mole of acid produces three moles of H^+ ions. Since three moles of

OH^- are required to react with the three moles of H^+ produced from one mole of the acid (see equation 1) we write for three moles of NaOH

$$3Na^+(aq) + 3OH^-(aq) \qquad (2)$$

The equation for the reaction between H^+ (from 1) and OH^- (from 2) is

$$3H^+(aq) + 3OH^-(aq) \rightarrow 3H_2O(l) \qquad (3)$$

Addition of (1), (2), and (3) then shows that the balanced equation for the required acid–base reaction is

$$KHC_2O_4 \cdot H_2C_2O_4 + 3NaOH \rightarrow K^+(aq) + 2C_2O_4^{2-}(aq) + 3Na^+(aq) + 3H_2O(l) \; (4)$$

(b) Even though Na_2CO_3 does not contain OH^- ions it is a base in aqueous solution because OH^- ions are formed by reaction with water.

$$Na_2CO_3 + 2H_2O \rightleftharpoons 2Na^+(aq) + H_2CO_3(aq) + 2OH^-(aq) \quad (5) \text{ (see note 1)}$$

The equation shows that one mole of Na_2CO_3 can give two moles of OH^- ions.

Two moles of H^+ ions, and therefore two moles of HCl are thus required to react with one mole of Na_2CO_3.

$$2HCl \rightarrow 2H^+(aq) + 2Cl^-(aq) \qquad (6)$$

The reaction between OH^- ions formed by reaction (5) and H^+ ions is

$$2H^+(aq) + 2OH^-(aq) \rightarrow 2H_2O \qquad (7)$$

Addition of equations (5), (6), and (7) shows that the balanced equation for the required reaction is (see note 2 below)

$$Na_2CO_3 + 2HCl \rightarrow 2Na^+(aq) + 2Cl^-(aq) + H_2O + CO_2 \qquad (8)$$

Notes

1 Reaction (5) is an equilibrium reaction and the forward reaction occurs only to a small extent. In an aqueous solution of Na_2CO_3 the concentration of OH^- ions is therefore low. When an acid is added to a sodium carbonate solution, the OH^- ions would be removed (because they react with H^+ ions). More of the forward reaction would then take place (see Le Chatelier's principle, Chapter 7). As more and more acid is added to a sodium carbonate solution, the forward reaction (see equation 5) would take place to a greater and greater extent. If sufficient acid is added the forward reaction would proceed to completion. A mole of Na_2CO_3 would then have produced two moles of OH^- ions. A similar argument applies to the equilibrium reaction (1).

2 In equation (8), CO_2 and H_2O are given as products instead of H_2CO_3. This is because H_2CO_3 would decompose into H_2O and CO_2.

Exercise 4.6

0.100 mol of sulphuric acid (H_2SO_4) is made up to 1.00 dm^3 of aqueous solution. How many grams of sodium carbonate (Na_2CO_3) must be added to 100 cm^3 of this solution just to neutralize all the acid?

Exercise 4.7

'Borax' has the formula $Na_2B_4O_7 \cdot 10H_2O$. It is a base in aqueous solution because the $B_4O_7^{2-}$ ions react with water to produce OH^- ions according to the equation

$$B_4O_7^{2-} + 7H_2O \rightleftharpoons 4H_3BO_3 + 2OH^-$$

How many grams of borax are required to react with 1.000 mol of hydrogen ions in hydrochloric acid?

4.4 Balancing oxidation–reduction reactions

A large number of chemical reactions are of the type known as oxidation–reduction (redox) reactions. In these one substance is oxidized while another substance is reduced.

Any oxidation–reduction reaction may be considered to be the sum of two half-reactions—one half-reaction is an oxdation reaction while the second half-reaction is a reduction reaction.

In an oxidation half-reaction electrons are released e.g.

$$Fe^{2+} \rightarrow Fe^{3+} + e^- \tag{1}$$

whereas in a reduction half-reaction electrons are accepted e.g.

$$Sn^{4+} + 2e^- \rightarrow Sn^{2+} \tag{2}$$

The balanced equation for any oxidation–reduction reaction could therefore be written by

(a) splitting up the reaction into two half-reactions
(b) writing the balanced equations for the two half-reactions
(c) adding together the two balanced equations for the half-reactions.

The method is illustrated by equation (3) below and in Example 4.5. A combination of the two half-reactions (1) and (2) leads to the complete redox equation by the elimination of e^-

$$2Fe^{2+} + Sn^{4+} \rightarrow 2Fe^{3+} + Sn^{2+} \tag{3}$$

Equation (3) is consistent with the laws of conservation of mass and of charge.

Example 4.5 Writing the balanced equation for an oxidation–reduction reaction

Write the balanced equation for the oxidation of Fe^{2+} by MnO_4^- in an aqueous acid medium.

Solution

To balance any equation we must first know the reactants and products; these are experimental quantities. The products in the reaction considered are Fe^{3+} and Mn^{2+}. The equation (unbalanced) for the reaction is therefore

$$MnO_4^-(aq) + Fe^{2+}(aq) \rightarrow Mn^{2+}(aq) + Fe^{3+}(aq) \tag{1}$$

The problem is to balance this equation. As stated already, balancing can be done by splitting up the reaction into two half-reactions, balancing each half-reaction, and then adding the balanced half-reactions together.

The half-reactions involved in reaction (1) are

$$Fe^{2+}(aq) \rightarrow Fe^{3+}(aq) \qquad (2)$$

$$MnO_4^-(aq) \rightarrow Mn^{2+}(aq) \qquad (3)$$

To balance these half-reactions we have to apply the law of conservation of matter (i.e. of atoms) and the law of conservation of charge.

In equation (2) (the oxidation half-reaction) the law of conservation of matter is already satisfied. To satisfy the law of conservation of charge we must rewrite equation (2) as follows

$$Fe^{2+}(aq) \rightarrow Fe^{3+}(aq) + e^- \qquad (4)$$

In equation (3) (the reduction half-reaction) Mn is balanced but O is not. There are four atoms of O on the left side of the equation; so there should be four atoms of O on the right side also. The question arises: 'What happens to the O in MnO_4^- during reaction?' As a general rule, the O in any oxidizing agent reacts with H^+ ion, (remember that all aqueous solutions contain H^+ ions) to form H_2O. The four O atoms in MnO_4^- would therefore react with $8H^+$ to form $4H_2O$. The reduction half-reaction (3) could therefore be rewritten as

$$MnO_4^-(aq) + 8H^+(aq) \rightarrow Mn^{2+}(aq) + 4H_2O(l) \qquad (5)$$

The charge, however, is not balanced. The left side has a net charge $+7$ while the right side has a charge $+2$. To balance charges, we add the required number of electrons. The equation then becomes

$$MnO_4^-(aq) + 8H^+(aq) + 5e^- \rightarrow Mn^{2+}(aq) + 4H_2O(l) \qquad (6)$$

Equations (4) and (6) are the balanced equations for the two half-reactions. Equation (4) involves one electron but equation (6) involves five electrons. In any oxidation–reduction reaction the electrons released during the oxidation half-reaction are absorbed by the reduction half-reaction. The same number of electrons must therefore be involved in each half-reaction. We therefore multiply equation (4) by 5 when we obtain

$$5Fe^{2+}(aq) \rightarrow 5Fe^{3+}(aq) + 5e^- \qquad (7)$$

By adding equations (6) and (7) we obtain the balanced equation for the required oxidation–reduction

$$MnO_4^-(aq) + 8H^+(aq) + 5Fe^{2+}(aq) \rightarrow Mn^{2+}(aq) + 4H_2O(l) + 5Fe^{3+}(aq) \qquad (8)$$

As a check we could work in oxidation numbers

$$Fe(III) \rightarrow Fe(II) + e^- \qquad (2a)$$

$$Mn(VII) + 5e^- \rightarrow Mn(II) \qquad (3a)$$

and combining (2a) and (3a) we obtain

$$Mn(VII) + 5Fe(II) \rightarrow Mn(II) + 5Fe(III) \qquad (8a)$$

We also see that (8) is consistent with the laws of conservation of mass and of charge.

Exercise 4.8

Iodine (I_2) reacts with thiosulphate ($S_2O_3^{2-}$) ions to give iodide (I^-) ions and tetrathionate ($S_4O_6^{2-}$) ions. Write the balanced equations for the two half-reactions.

Exercise 4.9

Write the balanced equation for the oxidation of ethanedioate ($C_2O_4^{2-}$) to carbon dioxide by dichromate(VI) ($Cr_2O_7^{2-}$). The $Cr_2O_7^{2-}$ ions are then reduced to Cr^{3+}.

How many grams of $K_2Cr_2O_7$ are required to oxidize completely 1.000 g of $Na_2C_2O_4$?

Exercise 4.10

Combustion reactions in the gas phase may also be considered to be oxidation–reduction reactions. Balance the reaction involving the combustion of octane (C_8H_{18}) in oxygen by considering it to be an oxidation–reduction reaction.

Exercise 4.11

When MnO_2 reacts in an acid medium Mn^{2+} ion are formed. What is wrong with the following equation for this half-reaction?

$$MnO_2(s) + 4H^+(aq) \rightarrow Mn^{2+}(aq) + 2H_2O(l) + 2e^-$$

4.5 Ways of expressing concentrations

So far, balanced equations of reactions were interpreted in terms of amounts (moles). Often, particularly when dealing with reactions in solution (volumetric analysis, gravimetric analysis, ionic equilibria, etc.) it is found more convenient to deal with *concentrations* (rather than amounts) of reactants and products.

Concentrations have been expressed in various ways. The most important are: percentage by mass, concentration*, molality, and mole fraction.

In this section we first state the definitions of these concentration terms; then we illustrate some calculations involving them. In particular, we show how calculations based on a balanced equation can be done when data are given in concentrations.

Percentage (%)

The percentage by mass (see section 2.2 and **Equation 2.3**) of a substance B in a system is defined by the equation

$$\% \text{ B (by mass)} = \frac{m_B}{m} \times 100 \qquad \textbf{Equation 4.1}$$

where m_B = mass of B and m = total mass of the system.

Concentration* (c)

Concentration (c_B) of a substance B in a system is defined as the moles of B (n_B) present per unit volume of the system. That is

* Formerly known as molar concentration. (See also footnotes on pages 62 and 215.)

$$c_B = \frac{n_B}{V} \qquad \text{Equation 4.2}$$

The SI unit for volume (see section 2.4) is metre^3 (m^3) and therefore the SI unit for molar concentration is mol/m^3 or mol m^{-3} but in chemistry mol dm^{-3} is nearly always used instead.

Molality* (m)

Molality (m_B) of a substance B in a system is defined as the moles n_B of B present per unit mass of *solvent*. That is

$$m_B = \frac{n_B}{m_{solvent}} \qquad \text{Equation 4.3}$$

where $m_{solvent}$ is the mass of solvent dissolving n_B mol of B.

Mole fraction (x)

Mole fraction (x_B) of a substance B in a system is defined as the moles n_B of B divided by the total moles (n) of all the substances in the system. That is

$$x_B = \frac{n_B}{n} \qquad \text{Equation 4.4}$$

Mole fraction $\times$ 100 is known as mole percent (mol%).

Example 4.6 Calculation of concentration

6.124 g of $CuSO_4 \cdot 5H_2O$ are dissolved in water and the solution is made up to 250.0 cm^3 in a volumetric flask. Calculate the concentration, in SI units, of the salt in the solution.

Solution

Step 1 Clarify and define the problem
We have to calculate the concentration c of $CuSO_4 \cdot 5H_2O$ in the solution.

Step 2 Select the key equation
As recommended in Chapter 1 we start with the defining equation for the required quantity c which is

$$c = \frac{n}{V} \qquad \text{(1) Equation 4.2}$$

n is the moles of the solute present in a volume V of the solution.

* Formerly called molar concentration.

Step 3   Derive the equation for the calculation

Equation (1) shows that to calculate c we must know n and V. The value for V is given; that for n is not. We must therefore try to relate n to the given data; this can be done by the equation

$$n = \frac{m}{M} \qquad\qquad (2) \textbf{ Equation 3.3}$$

where m and M are respectively the mass and the molar mass of the solute.
On replacing n in equation (1) by equation (2) one obtains

$$c = \frac{m}{MV} \qquad\qquad (3)$$

Equation (3) can be used to calculate the required molar concentration since all the quantities required for its calculation are known.

Step 4   Collect the data, check the units, and calculate
It is given that

$$m = 6.124 \text{ g}$$

$$V = 250.0 \text{ cm}^3$$

To calculate M for $CuSO_4 \cdot 5H_2O$ we require A_r values for Cu, S, O, and H; these can be looked up from a table (see page 213)

$$M \text{ (for } CuSO_4 \cdot 5H_2O) = (63.54 + 32.06) + (4 \times 16.00) + (5 \times 18.016) \text{ g mol}^{-1}$$

$$= 249.7 \text{ g mol}^{-1}$$

$$\therefore \quad c = \frac{6.124 \text{ g}}{(249.7 \text{ g mol}^{-1})(250.0 \text{ cm}^3)}$$

$$= 9.810 \times 10^{-5} \text{ mol cm}^{-3} \qquad\qquad (4)$$

Step 5   Review and check the solution
The calculation above shows that $c = 9.810 \times 10^{-5}$ mol cm^{-3}. On re-reading the problem statement it is, however, seen that what is required is c in SI units (i.e. in mol m^{-3}). Since 1 cm $= 10^{-2}$ m it follows that (see section 2.1)

$$(1 \text{ cm})^3 = (10^{-2} \text{ m})^3 = 10^{-6} \text{ m}^3$$

$$\therefore \quad 1 \text{ cm}^{-3} = 10^6 \text{ m}^{-3} \qquad\qquad (5)$$

To express the concentration in mol m^{-3} units we simply replace cm^{-3} in equation (4) by equation (5). We then obtain

$$c = 9.810 \times 10^{-5} \text{ mol} \times 10^6 \text{ m}^{-3}$$

$$c = 98.1 \text{ mol m}^{-3}$$

Exercise 4.12

How many grams of potassium dichromate(VI) ($K_2Cr_2O_7$) must be dissolved in 250.0 cm³ of water to prepare a 0.6798 mol dm⁻³ solution?

Example 4.7 Calculation of concentration from percentage by mass data

Calculate the concentration of a solute B in a solution which contains 2.00% by mass of B. The density of the solution is 1.145 g cm⁻³ and the molar mass of B is 246.2 g mol⁻¹.

Solution

Step 1 Clarify and define the problem

The problem is to calculate concentration (c) from percentage by mass data. To do so, we have to derive an equation that shows the relationship between c and % by mass.

Step 2 Select the key equation

We start with the defining equation for the required quantity—concentration c—which is

$$c = \frac{n}{V}$$

(1) **Equation 4.2**

Step 3 Derive the equation for the calculation

We have to derive an equation relating c and % by mass. So we now write the defining equation for % by mass

$$\% \text{ B} = \frac{m_B}{m_{solution}} \times 100$$

(2) **Equation 4.1**

where % B is the percentage of B (by mass) in the solution, m_B is the mass of B in the solution, and $m_{solution}$ is the mass of the solution.

The relationship between the molar concentration (c) of B in the solution and % B could be obtained by combining equations (1) and (2) together. If, for example, we divide equation (1) by equation (2) we obtain

$$\frac{c}{\% \text{ B}} = \frac{n}{V} \frac{m_{solution}}{m_B \times 100}$$

(3)

Equation (3) shows that to determine c from % B we have to know the values of all the quantities on the right side of the equation. Are these obtainable from the data? Yes, because ρ, density of solution, by definition is

$$\frac{m_{solution}}{V} = \rho$$

(4)

and by **Equation 3.3** we know that n/m_B is related to the molar mass M.

$$\frac{n}{m_B} = \frac{1}{M}$$

(5)

On inserting equations (4) and (5) in equation (3) and on rearranging we obtain the equation required for the calculation

$$c = \frac{\% \text{ B} \times \rho}{M \times 100} \qquad (6)$$

Step 4 Collect the data, check the units, and calculate

From data

$$\% \text{ B} = 2.000$$

$$\rho = 1.145 \text{ g cm}^{-3}$$

$$M = 246.2 \text{ g mol}^{-1}$$

By equation (6) we therefore have

$$c = \frac{2.00 \times 1.145 \text{ g cm}^{-3}}{(246.2 \text{ g mol}^{-1}) \times 100}$$

$$= 9.30 \times 10^{-5} \text{ mol cm}^{-3} = 9.30 \times 10^{-5} \text{ mol } (10^6 \text{ m}^{-3})$$

$$= 93.0 \text{ mol m}^{-3}$$

Step 5 Review and check the solution

The units in the answer are correct and the number of significant figures is appropriate.

Example 4.8 Calculation involving concentration

A sample of commercial concentrated hydrochloric acid, which is an aqueous solution, contains 36.0% by mass of hydrogen chloride and has a density 1.18 g cm^{-3}. Indicate how you would prepare, starting with this acid, 0.800 dm^3 of a 0.200 mol dm^{-3} solution of hydrochloric acid. Molar mass of HCl is 36.46 g mol^{-1}.

Solution

Step 1 Clarify and define the problem

The question we have to answer can be stated more precisely as follows: 'What mass of the given concentrated hydrochloric acid solution must we make use of in order to prepare, by dilution with water, 0.800 dm^3 of a 0.200 mol dm^{-3} solution of hydrochloric acid?' Let $m_{solution}$ be this mass.

To recognize the method for finding $m_{solution}$ we should first identify the *principle* involved in this type of calculation. The basic principle involved here is that m_{HCl} (i.e. the mass of HCl) in the amount of concentrated solution that we take should be equal to the mass of HCl in the solution we have to prepare. Two solutions are involved in this problem (the solution of concentrated hydrochloric acid and the solution we have to prepare) and it would be clearer if we consider each solution separately and do the calculation in two parts.

(a) First calculate the mass of HCl that is present in the solution we have to prepare (0.800 dm^3 of 0.200 mol dm^{-3} HCl solution).

(b) Then calculate m_{solution}—the mass of the solution of concentrated hydrochloric acid that must be taken to prepare the required solution.

We first consider calculation (a).

Calculation (a)

Step 2　Select the key equation

The mass of HCl (m_{HCl}) present in a solution is given by

$$m_{\text{HCl}} = n \times M \qquad \text{(1)} \; \textbf{Equation 3.3}$$

where n is the moles of HCl present in the solution and M is the molar mass of HCl.

Step 3　Derive the equation for calculation

To calculate m_{HCl} by equation (1) we require n; this is not given in the data. n is, however, related to the given data by

$$n = c \times V \qquad \text{(2)} \; \textbf{Equation 4.2}$$

On replacing n in equation (1) by equation (2) we obtain

$$m_{\text{HCl}} = c \times V \times M \qquad \text{(3)}$$

The mass of HCl (m_{HCl}) present in 0.800 dm^3 of a 0.200 mol dm^{-3} solution of HCl can be calculated by the application of equation (3).

Step 4　Collect the data, check the units, and calculate

$$c = 0.200 \text{ mol dm}^{-3}$$

$$V = 0.800 \text{ dm}^3$$

$$M = 36.46 \text{ g mol}^{-1}$$

∴ $$m_{\text{HCl}} = (0.200 \text{ mol dm}^{-3})\,(0.800 \text{ dm}^3)\,(36.46 \text{ g mol}^{-1})$$

$$= 5.834 \text{ g}$$

Calculation (b)

Calculation (a) shows that 5.834 g of HCl would be present in the solution we have to prepare. The mass of the concentrated HCl solution (m_{solution}) that we take for dilution should therefore also contain 5.834 g HCl. Since the concentrated HCl solution has 36.0% by mass of HCl, it follows by the application of **Equation 4.1** that

$$m_{\text{solution}} = 5.834 \text{ g} \times \frac{100}{36.0}$$

$$= 16.2 \text{ g}$$

That is, to obtain 0.800 dm^3 of a 0.200 mol dm^{-3} solution of HCl, 16.21 g of the given concentrated HCl solution must be diluted to 0.800 dm^3 with distilled water.

Step 5　Review and check the solution

Exercise 4.13

Ammonia is sold commercially as a concentrated aqueous solution. The specifications given in a bottle containing concentrated ammonia solution are: density 0.91 g cm^{-3}; % NH_3 by weight 25.0. What is the molar concentration of ammonia in this solution? Starting with this solution how would you prepare 1.00 dm^3 of a 1.00 mol dm^{-3} solution of ammonia?

Exercise 4.14

From a 0.1933 mol dm^{-3} solution of sulphuric acid, how would you prepare 1000 cm^3 of a 0.1000 mol dm^{-3} solution?

Example 4.9 Calculation of molality from concentration

Calculate the molality of a solute in a solution whose concentration is 0.500 mol dm^{-3}. The density of the solution is 0.988 kg dm^{-3} and the molar mass of the solute is 42.0 g mol^{-1}.

Solution

Step 1 Clarify and define the problem
The problem is to calculate molality (m) from concentration (c). We have therefore to derive an equation which shows the relationship between m and c.

Step 2 Select the key equation
We start with the defining equation for the required quantity which is

$$m = \frac{n_{solute}}{m_{solvent}} \qquad \text{(1) Equation 4.3}$$

Step 3 Derive the equation for the calculation
n_{solute} is related to the given data (c) by the equation

$$n_{solute} = c \times V \qquad \text{(2) Equation 4.2}$$

On replacing n_{solute} in equation (1) by equation (2) we obtain

$$m = \frac{cV}{m_{solvent}} \qquad \text{(3)}$$

Equation (3) shows that to obtain m from c we must know the mass of solvent ($m_{solvent}$) present in a known volume of solution (V); this can be calculated as follows:
    Consider 1.000 dm^3 of solution. It has a mass 0.988 kg (since mass = volume × density = 1.000 dm^3 × 0.988 kg). Since the mass of a solution is equal to the sum of the masses of the solute and of the solvent (from the principle of *additivity* of masses) it follows that

$$m_{solvent} = m_{solution} - m_{solute}$$
$$= 0.988 \text{ kg} - 0.021 \text{ kg} \qquad \text{(see note below)}$$
$$= 0.967 \text{ kg}$$

That is, 0.967 kg of solvent is present in 1.000 dm³ of the solution. Equation (3) can now be used for the calculation of m since all the quantities required for its calculation are known.

Step 4 Collect the data, check the units, and calculate

$$c = 0.500 \text{ mol dm}^{-3}$$

$$V = 1.000 \text{ dm}^3$$

$$m_{\text{solvent}} = 0.976 \text{ kg}$$

$$\therefore \quad m = \frac{(0.500 \text{ mol dm}^{-3})\,(1.000 \text{ dm}^3)}{0.967 \text{ kg}}$$

$$= 0.517 \text{ mol kg}^{-1}$$

Step 5 Review and check the solution

The answer is reasonable, the units are correct, and the number of significant figures is appropriate.

Note

The mass of solute present in 1.000 dm³ of a 0.500 mol dm⁻³ solution is $(42.0/2)$ g $=$ 0.021 kg, since the molar mass of solute is 42.0 g mol⁻¹—check this as an exercise.

Exercise 4.15

10.00 g of a substance of molar mass 150.0 g mol⁻¹ are dissolved in water and the solution is made up to 100.0 cm³ at 25 °C. If the density of the solution obtained is 1.045 g cm⁻³, calculate the molality of the solution.

Exercise 4.16

The molality of a solution containing 10.00 g of a solute in 200.0 g of water is 0.6270 mol kg⁻¹. Calculate the molar mass of the solute.

Example 4.10 Concentrations and mole fractions after gaseous dissociation

On strong heating, ammonia dissociates into nitrogen and hydrogen. When 0.350 mol of NH_3 is heated in a closed 10.0 dm³ vessel the degree of dissociation is 0.400. Calculate the concentration and the mole fraction of each gas in the vessel.

Solution

Step 1 Clarify and define the problem

Since NH_3 dissociates partially into N_2 and H_2 it is clear that the reaction vessel contains NH_3, N_2, and H_2. We have therefore to calculate the concentrations (c) and mole fractions (x) of NH_3, N_2, and H_2. That is c_{NH_3}, c_{N_2}, c_{H_2}, x_{NH_3}, x_{N_2}, and x_{H_2} have to be calculated.

Step 2 Select the key equation

We start, as explained in Chapter 1, with the defining equation for the required quantity. Two physical quantities (c and x) have to be calculated; the two defining equations for a substance i (i $= NH_3$, N_2, or H_2) are

$$c_i = \frac{n_i}{V} \qquad\qquad \text{(1) Equation 4.2}$$

$$x_i = \frac{n_i}{n} \qquad\qquad \text{(2) Equation 4.4}$$

n_i is the moles of gas i and n is the total moles of all the gases present in the vessel.

Step 3 Derive the equation for the calculation

Equations (1) and (2) can be used to calculate c_i and x_i since V is known and n_i and n can be calculated by making use of the balanced equation for the dissociation which is

$$2NH_3 \rightleftharpoons N_2 + 3H_2$$

Step 4 Collect the data, check the units, and calculate

The data given and steps in the calculation are best tabulated as in Table 4.2.

Table 4.2 Data and calculations for Example 4.10

Balanced equation	$2NH_3$	$\rightleftharpoons$	N_2	$+$	$3H_2$
moles initially	0.350 mol		0		0
moles dissociated (see note 1)	(0.350×0.400) = 0.140 mol				
moles formed (see note 2)			0.070 mol		0.210 mol
moles present after dissociation, n_i	$(0.350 - 0.140)$ = 0.210 mol		0.070 mol		0.210 mol
concentrations (by equation 1)	$\left(\dfrac{0.210 \text{ mol}}{10.0 \text{ dm}^3}\right)$ $= 2.10 \times 10^{-2}$ mol dm^{-3}		$\left(\dfrac{0.070 \text{ mol}}{10.0 \text{ dm}^3}\right)$ $= 7.0 \times 10^{-3}$ mol dm^{-3}		$\left(\dfrac{0.210 \text{ mol}}{10.0 \text{ dm}^3}\right)$ $= 210 \times 10^{-2}$ mol dm^{-3}
mole fractions (by equation 2; see also note 3)	$\left(\dfrac{0.210}{0.490}\right) = 0.429$		$\left(\dfrac{0.070}{0.490}\right) = 0.143$		$\left(\dfrac{0.210}{0.490}\right) = 0.429$

The concentrations of NH_3, N_2, and H_2 in the vessel are therefore 2.10×10^{-2} mol dm^{-3}, 7.10×10^{-3} mol dm^{-3}, and 2.10×10^{-2} mol dm^{-3} respectively. The mole fractions of NH_3, N_2, are 0.429, 0.143, and 0.429 respectively.

Step 5 Review and check the solution

Notes

1 The degree (or fraction) of dissociation (α) is defined by the equation

$$\alpha = \frac{\text{moles dissociated}}{\text{moles present initially}} \qquad \textbf{Equation 2.2}$$

$\therefore$ moles dissociated = moles present initially $\times$ α

$$= 0.350 \times 0.400$$

2 The moles of N_2 and H_2 formed by dissociation are calculated by making use of the balanced equation which shows that 2 moles of NH_3 gives, on complete dissociation, 1 mole of N_2 and 3 moles of H_2.

3 Total moles after dissociation $= n_{NH_3} + n_{N_2} + n_{H_2}$

$$= (0.210 + 0.070 + 0.210) \text{ mol}$$

$$= 0.490 \text{ mol}$$

Exercise 4.17

A gas A dissociates on heating into two other gases B and C according to the equation $A \rightarrow 2B + 3C$. When 0.050 mol of A is maintained at 100 °C in a closed 10.0 dm^3 vessel, the degree of dissociation is 0.30. Calculate (a) the mole fractions and (b) the concentrations, of A, B, and C in the vessel.

Exercise 4.18

When 0.500 mol of N_2 and 1.000 mol of H_2 are mixed in a 5.00 dm^3 vessel, 3.406 g of NH_3 are formed. Calculate the mole fractions of N_2, H_2, and NH_3 in the vessel.

Example 4.11 Concentrations after ionic dissociation

In aqueous solution, sulphuric acid (H_2SO_4) dissociates in two steps according to the equations

$$H_2SO_4(aq) \rightarrow H^+(aq) + HSO_4^-(aq) \qquad (1)$$

$$HSO_4^-(aq) \rightleftharpoons H^+(aq) + SO_4^{2-}(aq) \qquad (2)$$

The first dissociation is complete while the second takes place only partially. In a 0.100 mol dm^{-3} solution of sulphuric acid the concentration of H^+ ions is 0.192 mol dm^{-3}. Calculate the concentration of SO_4^{2-} ions in the solution.

Solution

Step 1 Clarify and define the problem

We have to calculate the concentration of SO_4^{2-} ions in a solution. Let us represent this by $[SO_4^{2-}]$.*

---

* Square brackets [ ] are often used to denote concentrations. The symbol $[SO_4^{2-}]$ should be read as 'the concentration of SO_4^{2-} ions in mol dm^{-3}'.

Step 2 Select the key equation

SO_4^{2-} ions appear only in the dissociation equation (2) of the problem statement. As key equation we therefore select this equation

$$HSO_4^-(aq) \rightleftharpoons H^+(aq) + SO_4^{2-}(aq) \tag{3}$$

Step 3 Derive the equation for the calculation

Equation (2) shows that

$$[SO_4^{2-}] = [H^+]_2 \tag{4}$$

where $[H^+]_2$ is the concentration of H^+ ions produced only by dissociation (2). The concentration of H^+ ions in the solution, $[H^+]$, is the total hydrogen ion concentration produced by dissociations (1) and (2). That is

$$[H^+] = [H^+]_1 + [H^+]_2 \tag{5}$$

where $[H^+]_1$ is the hydrogen ion concentration produced by dissociation (1). On rearranging equation (5) we obtain

$$[H^+]_2 = [H^+] - [H^+]_1 \tag{6}$$

On replacing $[H^+]_2$ in equation (4) by equation (6) one obtains

$$[SO_4^{2-}] = [H^+] - [H^+]_1 \tag{7}$$

$[H^+]$ is given in the data (0.192 mol dm^{-3}) and therefore to calculate $[SO_4^{2-}]$ by equation (7) we must know $[H^+]_1$. Since dissociation (1) is complete it follows that

$$[H^+]_1 = c \tag{8}$$

where c is the 'initial' concentration of H_2SO_4 and is given in the data.

From equations (7) and (8) we obtain

$$[SO_4^{2-}] = [H^+] - c \tag{9}$$

Step 4 Collect the data, check the units, and calculate

From the data

$$[H^+] = 0.192 \text{ mol dm}^{-3}$$

$$c = 0.100 \text{ mol dm}^{-3}$$

By equation (9) we therefore have

$$[SO_4^{2-}] = 0.192 \text{ mol dm}^{-3} - 0.100 \text{ mol dm}^{-3}$$

$$= 0.092 \text{ mol dm}^{-3}$$

The concentration of SO_4^{2-} ions in the solution is therefore 0.092 mol dm^{-3}.

Step 5 Review and check the solution

Exercise 4.19

In a 0.100 mol dm^{-3} solution of a weak dibasic acid H_2A the concentrations of HA^- and A^{2-}

are 0.024 mol dm^{-3} and 0.004 mol dm^{-3} respectively. Calculate the concentrations of H$^+$ ions and H$_2$A molecules in the solution.

Exercise 4.20

In a solution of a weak dibasic acid H$_2$A the concentrations of H$^+$ and A^{2-} are 0.032 mol dm^{-3} and 0.004 mol dm^{-3} respectively. By making use only of the law of conservation of charge calculate the concentration of HA$^-$.

Exercise 4.21

Write the equation which shows the relationship between moles, concentration, and volume. Use this equation to calculate the moles of solute present in 25.00 cm^3 of a 0.1000 mol dm^{-3} solution.

Exercise 4.22

15.00 g of a substance of molar mass 150.0 g mol^{-1} are dissolved in water, and the solution is made up to 100.0 cm^3 at 25 °C. The density of this solution is 1.045 g cm^{-3} at 25 °C and 1.015 g cm^{-3} at 50 °C. Calculate (a) the concentrations (b) the molalities, at 25 °C and 50 °C.

Exercise 4.23

Calculate the mole fraction of CH$_3$COCH$_3$ in a solution obtained by dissolving 10.00 cm^3 pure CH$_3$COCH$_3$ at 25 °C in 50.00 g of water. The density of pure CH$_3$COCH$_3$ at 25 °C is 0.792 g cm^{-3}.

Exercise 4.24

5.00 g of a solute are dissolved in 100.0 g water. The mole fraction of the solute is 4.64 × 10^{-3}. Calculate the molar mass of the solute.

4.6 Volumetric and gravimetric analysis

Two important general methods of quantitative chemical analysis are volumetric analysis and gravimetric analysis. Calculations in volumetric analysis (where volumes are measured) and in gravimetric analysis (where masses are measured) are based on the balanced equation. Some examples are illustrated below.

Example 4.12 Calculations in volumetric analysis

Oxalate* ions (C$_2$O$_4^{2-}$) are oxidized to carbon dioxide by manganate(VII) ions (MnO$_4^-$) in an acid medium, the MnO$_4^-$ ions being reduced to manganese(II) ions (Mn^{2+}).

   Calculate the volume of a 0.1000 mol dm^{-3} solution of MnO$_4^-$ ions that would be required for the titration of 25.00 cm^3 of a 0.1234 mol dm^{-3} solution of C$_2$O$_4^{2-}$ ions.

   * *Systematic name is* ethanedioate.

Solution

Step 1 Clarify and define the problem

This calculation, like most calculations in volumetric analysis, is based on the information provided by the balanced equation for a reaction. The first step, therefore, is to write the balanced equation. The balanced equation (use the method given in Example 4.5 to write down this equation as an exercise) is

$$2MnO_4^- + 5C_2O_4^{2-} + 16H^+ \rightarrow 2Mn^{2+} + 10CO_2 + 8H_2O$$

In a titration we are concerned only with the reactants. In the problem given, the two reactants are MnO_4^- and $C_2O_4^{2-}$ and we need consider only these.

All relevant aspects connected with the problem are presented in Table 4.3; the subscripts 1 and 2 being used to refer to the MnO_4^- solution and the $C_2O_4^{2-}$ solution respectively.

Table 4.3 Presentation of the problem in Example 4.12

Balanced equation	$2MnO_4^-$	+	$5C_2O_4^{2-}$
	$c_1 = 0.1000$ mol dm^{-3} $V_1 = ?$		$c_2 = 0.1234$ mol dm^{-3} $V_2 = 25.00$ cm^3 $= 2.500 \times 10^{-2}$ dm^3

Step 2 Select the key equation

We require the volume V_1 of the MnO_4^- solution. This could be calculated by the application of the equation

$$V_1 = \frac{n_1}{c_1} \qquad \text{(1) \textbf{Equation 4.2}}$$

Step 3 Derive equation for calculation

It is seen that to calculate V_1 by equation (1) we require n_1 which is the moles of MnO_4^- required to react with the $C_2O_4^{2-}$ solution.

n_1 could be calculated from the data given by making use of the information provided by the balanced equation. The balanced equation shows that

$$n_1 = \frac{2}{5} \times n_2 \qquad (2)$$

where n_2 is the number of moles of $C_2O_4^{2-}$ present in 2.500×10^{-2} dm^3 of the 0.1234 mol dm^{-3} solution. n_2 is given by **Equation 4.2** as

$$n_2 = c_2 \times V_2 \qquad (3)$$

On combining equations (1), (2), and (3) we obtain

$$V_1 = \frac{(2/5) \times c_2 V_2}{c_1} \qquad (4)$$

Step 4 Collect the data, check the units, and calculate

$$c_2 = 0.1234 \text{ mol dm}^{-3}$$

$$V_2 = 2.500 \times 10^{-2} \text{ dm}^3$$

$$c_1 = 0.1000 \text{ mol dm}^{-3}$$

$$\therefore \quad V_1 = \frac{(2/5) \times (0.1234 \text{ mol dm}^{-3}) \, (2.500 \times 10^{-2} \text{ dm}^3)}{(0.1000 \text{ mol dm}^{-3})}$$

$$= 1.234 \times 10^{-2} \text{ dm}^3$$

$$= 12.34 \text{ cm}^3$$

Step 5 Review and check the solution

Exercise 4.25

Calculate the minimum volume of a 0.1000 M* acidified solution of $KMnO_4$ that would be required to oxidize completely (a) 1.000 g of $H_2C_2O_4$ (b) 25.00 cm³ of a 0.1000 M solution of a Fe^{2+} salt.

Exercise 4.26

When $KMnO_4$ acts as an oxidizing agent in an alkaline medium, MnO_4^- is reduced to MnO_2. Write the balanced equation for this half-reaction.

Calculate the moles of oxygen molecules (O_2) that would be evolved when excess H_2O_2 reacts with 25.0 cm³ of a 0.1000 M solution of $KMnO_4$ in an alkaline medium.

Example 4.13

Calculate the concentration of sodium hydroxide in the solution obtained when 10.00 cm³ of a 0.1000 mol dm⁻³ solution of sulphuric acid are added to 25.00 cm³ of a 0.0930 mol dm⁻³ solution of sodium hydroxide.

Solution

Step 1 Clarify and define the problem

The major difficulty in this problem is generally due to the fact that we have to keep track of three solutions—initial solutions of sodium hydroxide and sulphuric acid and the final solution obtained after mixing them.

Keeping track is made easier if we first write the balanced equation for the reaction taking place and then clearly present the given data in a systematic manner as follows:

$$2NaOH \quad + \quad H_2SO_4 \quad \rightarrow \quad Na_2SO_4 + 2H_2O \qquad (1)$$

$2NaOH$	H_2SO_4	$Na_2SO_4 + 2H_2O$
$V_1 = 25.00 \text{ cm}^3$	$V_2 = 10.00 \text{ cm}^3$	
$c_1 = 0.0930 \text{ mol dm}^{-3}$	$c_2 = 0.1000 \text{ mol dm}^{-3}$	
solution 1	solution 2	final solution

---

* The symbol M denotes concentrations in mol dm⁻³. This symbol is no longer recommended to be used, but Exercises 4.25, 4.26 and others have been included to give experience because many books still use M.

Step 2 Select the key equation

We want the molar concentration c of sodium hydroxide in the final solution. So we start with the defining equation for this quantity which is

$$c = \frac{n}{V} \qquad \text{(2) Equation 4.2}$$

n is the moles of sodium hydroxide present in the final solution and V is the volume.

Step 3 Derive the equation for the calculation

To calculate c by equation (2) we must know both n and V.

If we assume that volumes are additive (see note) it is clear that the volume V of the final solution is

$V = 25.00 \text{ cm}^3$ (from the NaOH solution) + 10.00 cm^3 (from the H_2SO_4 solution)

$= 35.00 \text{ cm}^3$

$= 3.50 \times 10^{-2} \text{ dm}^3$ \hfill (3)

Consider now the second term in equation (2): the moles n of NaOH present in the final solution. It is evidently given by

$n = $ mol NaOH in solution 1 $-$ mol NaOH reacted (when solution 2 is added) (4)

By Equation 4.2

$$\text{mol NaOH in solution } 1 = c_1 \times V_1 \qquad (5)$$

The balanced equation for the reaction (see equation 1) shows that

$$\text{mol NaOH reacted} = 2 \times \text{mol } H_2SO_4 \text{ present in solution 2}$$

$$= 2 \times c_2 V_2 \qquad (6)$$

On substituting equations (5) and (6) in equation (4) we then obtain

$$n = c_1 V_1 - 2 \times c_2 V_2 \qquad (7)$$

Combination of equations (2) and (7) then gives the required equation for the calculation

$$c = \frac{c_1 V_1 - 2c_2 V_2}{V} \qquad (8)$$

Step 4 Collect the data, check the units, and calculate

$c_1 = 0.0930 \text{ mol dm}^{-3}$

$V_1 = 25.00 \text{ cm}^3 = 2.500 \times 10^{-2} \text{ dm}^3$

$c_2 = 0.1000 \text{ mol dm}^{-3}$

$V_2 = 10.00 \text{ cm}^3 = 1.000 \times 10^{-2} \text{ dm}^3$

$V = 3.50 \times 10^{-2} \text{ dm}^3$ \hfill (see equation 3)

$$\therefore \quad c = \frac{\begin{array}{c}(0.0930 \text{ mol dm}^{-3} \times 2.500 \times 10^{-2} \text{ dm}^3) \\ - 2(0.1000 \text{ mol dm}^{-3} \times 1.000 \times 10^{-2} \text{ dm}^3)\end{array}}{3.50 \times 10^{-2} \text{ dm}^3}$$

$$= 9.29 \times 10^{-3} \text{ mol dm}^{-3}$$

Step 5 Review and check the solution

Note

In the calculation it was assumed that

$$25.00 \text{ cm}^3 \text{ NaOH solution} + 10.00 \text{ cm}^3 \text{ H}_2\text{SO}_4 \text{ solution} = 35.00 \text{ cm}^3$$

The error introduced by this assumption is negligibly small when dilute aqueous solutions are mixed with each other.

Exercise 4.27

How many grams of solid sodium hydroxide must be added to 100 cm³ of a 0.100 M solution of hydrochloric acid so that the concentration of the acid in the solution is reduced to 0.0100 M? Neglect the volume change due to the addition of sodium hydroxide.

Exercise 4.28

What volume of a 0.1000 M solution of sodium hydroxide must be added to 100.0 cm³ of a 0.1000 M solution of hydrochloric acid so that the concentration of HCl in the solution is 0.01000 M?

Example 4.14 Titration of iodine with sodium thiosulphate

A general method for the determination of the concentration of a solution of an oxidizing agent is by the addition of a known amount of this solution to a solution containing an excess of iodide. The iodine formed is then titrated with a standard solution of sodium thiosulphate.

25.00 cm³ of a potassium dichromate VI ($K_2Cr_2O_7$) solution were added to excess of acidified aqueous potassium iodide. The iodine then liberated required 19.63 cm³ of a 0.1478 mol dm⁻³ solution of sodium thiosulphate for complete reaction. Calculate the concentration of the potassium dichromate solution.

Solution

Step 1 Clarify and define the problem

It is seen that two reactions are involved—first the oxidation of I^- to I_2 by $Cr_2O_7^{2-}$ ions, and then the reduction of the I_2 formed by $S_2O_3^{2-}$ ions.

The balanced equations for the oxidation of I^- by $K_2Cr_2O_7$ and for the reduction of the I_2 by $Na_2S_2O_3$ are (use the method given in section 4.4 for writing these equations)

$$K_2Cr_2O_7 + 14H^+ + 6KI \rightarrow 8K^+ + 2Cr^{3+} + 3I_2 + 7H_2O \tag{1}$$

$$3I_2 + 6Na_2S_2O_3 \rightarrow 6I^- + 12Na^+ + 3S_4O_6^{2-} \tag{2}$$

The total change taking place, which could be obtained by adding together equations (1) and (2), is then

$$K_2Cr_2O_7 + 14H^+ + 6KI + 6Na_2S_2O_3 \rightarrow$$
$$8K^+ + 2Cr^{3+} + 7H_2O + 6I^- + 12Na^+ + 3S_4O_6^{2-} \qquad (3)$$

Equation (3) shows that one mole of $K_2Cr_2O_7$ requires six moles of $Na_2S_2O_3$. Now consider the problem given; the relevant information is presented together below (subscripts 1 and 2 refer to the $K_2Cr_2O_7$ solution and $Na_2S_2O_3$ solution respectively).

$K_2Cr_2O_7$ solution	$Na_2S_2O_3$ solution
$V_1 = 25.00$ cm^3	$V_2 = 19.63$ cm^3
$c_1 = ?$	$c_2 = 0.1478$ mol dm^{-3}

Step 2 Select the key equation
The required quantity—concentration c_1 of the $K_2Cr_2O_7$ solution—is defined by the equation

$$c_1 = \frac{n_1}{V_1} \qquad \text{(4) \textbf{Equation 4.2}}$$

Step 3 Derive the equation for the calculation
To calculate c_1 we require n_1 and V_1. Data give the volume V_1 of the $K_2Cr_2O_7$ solution; we have therefore to calculate only n_1. This could be done by making use of the balanced equation (3) and the data given.
    The balanced equation (3) shows that

$$6 \text{ mol } Na_2S_2O_3 \text{ is equivalent to } 1 \text{ mol } K_2Cr_2O_7$$

$$\therefore \qquad 1 \text{ mol } Na_2S_2O_3 \text{ is equivalent to } \tfrac{1}{6} \text{ mol } K_2Cr_2O_7$$

$$\therefore \qquad c_2 V_2 \text{ moles of } Na_2S_2O_3 \text{ would react with } \tfrac{1}{6} \times c_2 V_2 \text{ mol of } K_2Cr_2O_7$$

$$\text{i.e. } n_1 = \tfrac{1}{6} \times c_2 V_2 \qquad (5)$$

On substituting for n_1 in equation (4) by equation (5) we get the required equation for the calculation

$$c_1 = \frac{c_2 V_2}{6V_1}$$

Step 4 Collect the data, check the units, and calculate

$$c_2 = 0.1478 \text{ mol dm}^{-3}$$
$$V_2 = 19.63 \text{ cm}^3 = 1.963 \times 10^{-2} \text{ dm}^3$$
$$V_1 = 25.00 \text{ cm}^3 = 2.500 \times 10^{-2} \text{ dm}^3$$

$$\therefore \qquad c_1 = \frac{(0.1478 \text{ mol dm}^{-3}) \, (1.963 \times 10^{-2} \text{ dm}^3)}{6(2.500 \times 10^{-2} \text{ dm}^3)}$$

$$= 1.934 \times 10^{-2} \text{ mol dm}^{-3}$$

Step 5 Review and check the solution

Exercise 4.29

Borax has the formula $Na_2B_4O_7 \cdot 10H_2O$. It is a strong base in aqueous solution because OH^- ions are produced by reaction with water ($B_4O_7^{2-} + 7H_2O \rightleftharpoons 4H_3BO_3 + 2OH^-$). How many grams of borax are necessary (a) to prepare 100.0 cm³ of a 0.1000 M solution (b) just to neutralize 25.00 cm³ of a 0.1934 M solution of hydrochloric acid?

Exercise 4.30

What volume (cm³) of a 0.1965 M solution of sodium carbonate (Na_2CO_3) is necessary just to neutralize 100.0 cm³ of a 0.01965 M solution of ethanedioic acid ($H_2C_2O_4$)?

Exercise 4.31

What volume of a 0.1975 M solution of sodium hydroxide must be added to 100.0 cm³ of a 0.1975 M solution of hydrochloric acid so that the concentration of H^+ ions in the solution is 0.1000 M?

Exercise 4.32

Vegetable oils and fats essentially contain the glycerol esters of mixtures of long chain weak monobasic acids. When the oil becomes 'rancid' the esters partly hydrolyse to give the long chain acids (which are called fatty acids). One of the 'quality control' tests to check the purity of a sample of a fat or oil is the estimation of the amount of free fatty acids present by titration with an alkali.

10.00 cm³ of a sample of coconut oil required 29.8 cm³ of a 0.0500 M solution of sodium hydroxide for neutralization. Calculate the moles of H^+ (i.e. free acid) in the sample. If the average molar mass of the esters present in the oil is 250 g mol⁻¹ and the density of the oil is 0.91 g cm⁻³, calculate the percentage hydrolysis of the esters.

Exercise 4.33

A standard method for quantitatively estimating the amount of nitrogen in a substance is the *Kjeldahl method*. The nitrogen is converted into ammonia and the amount of ammonia is determined by titration.

The ammonia from 2.50 g of a foodstuff required 40.0 cm³ of a 0.0500 M solution of hydrochloric acid. Calculate the percentage of nitrogen in the foodstuff.

Exercise 4.34

25.00 cm³ of a 0.0500 M solution of sodium carbonate are required for the complete neutralization of a solution containing 0.2118 g of a tribasic acid. Calculate the molar mass of the acid.

Review of Chapter 4

Chapter 4 deals with some very important aspects of chemical equations—how to balance them, how to interpret them (in terms of molecules, moles, masses, concentrations, and volumes) and how to do calculations based on them. A thorough knowledge of this chapter is essential for understanding chemistry, and for solving problems in many sections of chemistry.

Recognize clearly and remember that

(a) the balancing of *all* types of equations is based on the same principles. Balancing is always done by the application of just two simple laws—the laws of conservation of matter and of charge.

(b) the calculations based on a balanced equation are, in principle, exactly similar for *all* types of reactions. All calculations involving a balanced equation depend basically on the fact that the equation indicates the number of moles of each reactant and product that participates in the reaction.

The moles (n) involved in a balanced equation should be converted, whenever required, into the number of molecules (N) by the equation $L = N/n$ (**Equation 3.2**), into masses (m) by the equation $M = m/n$ (**Equation 3.3**), into volumes (V) by the equation $V = nRT/P$ (**Equation 5.1**), into concentration (c) by **Equation 4.2**, into molality (m_B) by **Equation 4.3**, and into mole fraction (x_B) by **Equation 4.4**.

For the interpretation of equations in terms of concentrations, an understanding of some of the concentration concepts is required. Four simple equations involving concentrations, which you should be able to recall, understand, and apply, are

$$\% \text{ B (by mass)} = \frac{m_B}{m} \qquad \textbf{4.1}$$

$$\text{concentration, } c_B = \frac{n_B}{V} \qquad \textbf{4.2}$$

$$\text{molality, } m_B = \frac{n_B}{m_{\text{solvent}}} \qquad \textbf{4.3}$$

$$\text{mole fraction, } x_B = \frac{n_B}{n} \qquad \textbf{4.4}$$

5 | Gases

5.1 The ideal gas equation

An ideal gas is one in which there are no forces of attraction or repulsion between the molecules. For all ideal gases the relationship between pressure p, volume V, amount (moles) n, and temperature T is given by the equation

$$pV = nRT$$

Equation 5.1

R is known as the ideal gas constant and it has a value of 8.314 J K^{-1} mol^{-1}.

Gases obey the ideal gas equation accurately only at very low pressures. At atmospheric pressure, for example, all gases show deviations from ideal behaviour. These deviations, however, are generally not large. The ideal gas equation is therefore generally applied to describe the behaviour of all gases at normal pressures.

Analysis of equation $pV = nRT$

We have already emphasized (see section 2.2) that a clear understanding of how to extract the information summarized in an equation is essential for an understanding of the physical sciences. Let us therefore outline some of the information summarized in the ideal gas equation.

The equation $pV = nRT$ shows quantitatively how, for an ideal gas, the different variables (p, V, n, and T) depend on each other. The equation shows, for example, that

(a) at constant n and T, the product pV is a constant, or that p is inversely proportional to V (which is *Boyle's law*)
(b) at constant n and p, the volume V is directly proportional to T (which is *Charles' law*)
(c) at constant n and V, the pressure p is directly proportional to T
(d) at constant p and T, the volume V is directly proportional to n
(e) at constant V and T, the pressure p is directly proportional to n.

From the equation we could also deduce (because the constant R has the same value for *all* gases) that the physical behaviour of an ideal gas does not depend on the nature of the gas.

Example 5.1 Calculation of the ideal gas constant R

The volume of 1.000 mole of any gas, provided it behaves ideally, is found to be 22 414 cm^3 at 273.15 K and 1.000 atm pressure. Calculate the value of the gas constant R in SI units (1 atm = 1.0133 × 10^5 Pa).

Solution

Step 1 Clarify and define the problem

We have to calculate the value of the gas constant R in SI units.

Step 2 Select the key equation

To calculate R we must start with an equation which relates R to the given data. This equation is the ideal gas equation:

$$pV = nRT \qquad \textbf{Equation 5.1}$$

For convenience in calculation, we rearrange this equation so that only the required quantity R is on the left side; we then obtain:

$$R = \frac{pV}{nT} \qquad (1)$$

Step 3 Derive the equation for the calculation

All the quantities necessary for the calculation of R by the above equation are seen to be given in the data. The key equation can therefore be used directly for the calculation of R.

Step 4 Collect the data, check the units, and calculate

To calculate R in SI units, using equation (1), we have to give p, V, n, and T in SI units. On doing so we have

$$V = 2.2414 \times 10^{-2} \, \text{m}^3 \qquad \text{(see note 1)}$$

$$p = 1.0133 \times 10^5 \, \text{kg m}^{-1} \, \text{s}^{-2} \qquad \text{(see note 2)}$$

$$T = 273.15 \, \text{K}$$

$$n = 1.000 \, \text{mol}$$

We can now calculate R by making use of equation (1).

$$R = \frac{(1.0133 \times 10^5 \, \text{kg m}^{-1} \, \text{s}^{-2}) \, (2.2414 \times 10^{-2} \, \text{m}^3)}{(1.000 \, \text{mol}) \, (273.15 \, \text{K})}$$

$$= 8.314 \, \text{kg m}^2 \, \text{s}^{-2} \, \text{K}^{-1} \, \text{mol}^{-1}$$

$$= 8.314 \, \text{J K}^{-1} \, \text{mol}^{-1} \text{ (since 1 kg m}^2 \, \text{s}^{-2} = 1 \, \text{J, see note 1 below)}$$

Step 5 Review and check the solution

The units are correct and the numerical value can be checked against the known value (Appendix 1).

Notes

1 Read section 2.4 for an introduction to SI units.

The SI unit for volume (V) is cubic metre (m^3). Data give volume in cubic centimetres (cm^3). Since 1 cm $= 10^{-2}$ m it follows that

$$1 \, \text{cm}^3 = (10^{-2} \, \text{m})^3 = 10^{-6} \, \text{m}^3$$

The data given (in cm^3) could therefore be converted into the SI unit (m^3) by multiplying by 10^{-6}. That is

$$22\,414\,cm^3 = 22\,414\,(10^{-6}\,m^3) = 2.2414 \times 10^{-2}\,m^3$$

2 The pascal (symbol Pa) and the joule (symbol J) are special names for the SI units for pressure and energy respectively (see Table 2.4). You should be able to relate, whenever necessary, these units to the basic units (kg, m, s,...). This is easily done by first simplifying, step by step, the definition of the physical quantity. We illustrate this method by considering pressure as an example.

$$\text{pressure} = \frac{\text{force}}{\text{area}}$$

$$= \frac{\text{mass} \times \text{acceleration}}{(\text{length})^2}$$

$$= \frac{\text{mass}}{(\text{length})^2} \frac{\text{velocity change}}{\text{time}}$$

$$= \frac{\text{mass}}{(\text{length})^2 \times \text{time}} \frac{\text{length}}{\text{time}}$$

$$= \frac{\text{mass}}{\text{length} \times (\text{time})^2} \tag{2}$$

The above equation relates pressure to the basic physical quantities mass, length, and time.

To obtain the SI unit for pressure we merely have to insert the SI units for the physical quantities on the right side of equation (2), and then simplify. That is

$$\text{SI unit for pressure} = \frac{kg}{m \times s^2} = kg\,m^{-1}\,s^{-2}$$

$kg\,m^{-1}\,s^{-2}$, the SI unit for pressure, is given the special name pascal (Pa). That is

$$Pa = kg\,m^{-1}\,s^{-2} \tag{3}$$

Similarly it is easily seen that the SI unit for energy, the joule (J), is given by

$$J = kg\,m^2\,s^{-2} \tag{4}$$

Since Pa and J will occur frequently in your calculations, it may be worthwhile memorizing equations (3) and (4).

3 In this calculation, as well as in the other calculations in this book, the unit of each physical quantity (in terms of the basic units) has been inserted along with the numerical value. You should always remember to do this: some of the advantages of doing so were outlined in section 2.4.

Exercise 5.1

Use the data given in Example 5.1 and show that the gas constant R has the value 0.0821 dm^3 atm K^{-1} mol^{-1} if volume is expressed in dm^3 and pressure in atm.

Exercise 5.2

Suppose the amount of gas is given in terms of its mass m (instead of moles n). The relationship between p, V, m, and T for a particular gas may then be written as $pV = mkT$ where k is a constant. Is the constant k different from the gas constant R? Why is it better to give the amount of a gas in terms of moles, rather than in terms of mass?

Example 5.2   Application of the equation $pV = nRT$

A 1.00 dm³ vessel contains oxygen gas at 25 °C and at a pressure 4.05×10^5 Pa. Calculate the number of molecules of oxygen present in the vessel.

Solution

Step 1   Clarify and define the problem

We have to calculate the number of molecules N of oxygen present in the vessel.

Step 2   Select the key equation

The required quantity N is related to the Avogadro constant L by the equation

$$N = n \times L \qquad\qquad \text{(1) Equation 3.2}$$

Step 3   Derive the equation for the calculation

Equation (1) cannot be used for the calculation of N because n, the number of moles of gas present in the vessel, is not given in the data. We should therefore try to replace n by physical quantities given in the data. From the ideal gas equation it is clear that

$$n = \frac{pV}{RT} \qquad\qquad \text{(2) Equation 5.1}$$

On replacing n in equation (1) by equation (2) we get

$$N = \frac{pV}{RT} \times L \qquad\qquad \text{(3)}$$

Equation (3) can be used to calculate the required quantity N since all the quantities on the right side of the equation are known.

Step 4   Collect the data, check the units, and calculate

On expressing the data given in SI units we obtain

$$p = 4.05 \times 10^5 \text{ Pa} = 4.05 \times 10^5 \text{ kg m}^{-1}\text{s}^{-2}$$

$$V = 1.00 \times 10^{-3}\text{ m}^3$$

$$T = (273 + 25)\text{ K} = 298\text{ K}$$

Values for L and R are obtained from tables of physical constants.

$$L = 6.022 \times 10^{23}\text{ mol}^{-1}$$

$$R = 8.314 \text{ J K}^{-1}\text{ mol}^{-1} = 8.314 \text{ kg m}^2\text{ s}^{-2}\text{ K}^{-1}\text{ mol}^{-1}$$

$$\therefore \quad N = \frac{(4.05 \times 10^5 \text{ kg m}^{-1}\text{ s}^{-2})\,(1.00 \times 10^{-3}\text{ m}^3)}{(8.314 \text{ kg m}^2\text{ s}^{-2}\text{ K}^{-1}\text{ mol}^{-1})\,(298\text{ K})} \times (6.022 \times 10^{23}\text{ mol}^{-1})$$

$$= 9.84 \times 10^{22}$$

Step 5 Review and check the solution

Unless otherwise specified it should be assumed, as we have done in this example, that the equation $pV = nRT$ is applicable for calculations involving gases.

Students sometimes solve problems on gases by starting with a statement of the form, 'One mole of any gas occupies 22 414 cm^3 at 298 K and one atmosphere pressure'. This statement expresses in words some of the information contained in the equation $pV = nRT$. The remembering of information in the form of equations, and the use of equations for calculations, have many advantages (read section 1.4). We therefore strongly recommend that you use equations for calculations, rather than statements.

Example 5.3 Application of the equation $pV = nRT$

Gases such as nitrogen and oxygen are commonly used in the laboratory and they can be purchased in steel cylinders. A nitrogen cylinder of volume 100 dm^3 has a pressure 2.00×10^3 kPa at 27 °C. Nitrogen is removed from the cylinder until the pressure drops to 1.20×10^3 kPa. How many moles of nitrogen molecules were removed?

Solution

Step 1 Clarify and define the problem

We first clarify the problem and present it clearly. The problem is seen to involve two states: an initial state and a final state. The relevant data, converted into SI units, corresponding to these two states could be arranged in a convenient way as follows:

$V = 0.100$ m^3 $T = 300$ K		$V = 0.100$ m^3 $T = 300$ K	
$p_1 = 2.00 \times 10^6$ Pa		$p_2 = 1.20 \times 10^6$ Pa	
initial state		*final state*	

Place emphasis on the required physical quantity—the moles of nitrogen (N_2) removed. Evidently

$$\text{moles of N}_2 \text{ removed} = n_1 - n_2 \tag{1}$$

where n_1 and n_2 are the moles present in the initial and final state respectively.

Step 2 Select the key equation

n_1 and n_2 could be calculated by making use of the equation $pV = nRT$.

Application of this equation to the initial state shows that n_1 is given by

$$n_1 = \frac{p_1 V}{RT} \tag{2}$$

where p_1 = pressure in the initial state. A similar application of the ideal gas equation to the final state gives

$$n_2 = \frac{p_2 V}{RT} \tag{3}$$

where p_2 = pressure in the final state.

Step 3 Derive the equation for the calculation

We want $n_1 - n_2$ (see equation 1). This could be obtained by subtracting equation (3) from equation (2) when we obtain

$$n_1 - n_2 = \frac{(p_1 - p_2) V}{RT} \tag{4}$$

Equation (4) could be used for calculating the required quantity $(n_1 - n_2)$ since all the quantities required for its calculation are given, except R which is a physical constant (Appendix 1).

Step 4 Collect the data, check the units, and calculate

$$p_1 = 2.00 \times 10^6 \text{ Pa (kg m}^{-1} \text{ s}^{-2})$$

$$p_2 = 1.20 \times 10^6 \text{ Pa (kg m}^{-1} \text{ s}^{-2})$$

$$V = 1.00 \times 10^{-1} \text{ m}^3$$

$$T = 300 \text{ K}$$

$$n_1 - n_2 = \frac{(p_1 - p_2) V}{RT}$$

$$= \frac{(0.80 \times 10^6 \text{ kg m}^{-1} \text{ s}^{-2}) (1.00 \times 10^{-1} \text{ m}^3)}{(8.31 \text{ kg m}^2 \text{ s}^{-2} \text{ K}^{-1} \text{ mol}^{-1}) (300 \text{ K})}$$

$$= 32.1 \text{ mol}$$

Moles of nitrogen molecules removed are therefore 32.1 mol.

Step 5 Review and check the solution

In this example we could have calculated the values of n_1 and n_2 individually, and then used equation (1) to calculate the moles of nitrogen removed. This procedure, however, would have involved unnecessary arithmetical operations, and would therefore have been more time consuming.

You will find it advantageous to insert the values for the physical quantities and perform the calculation at the final step.

Example 5.4 Dependence of density on pressure

Show quantitatively how the density of an ideal gas would depend on the pressure.

Solution

Step 1 Clarify and define the problem

To deduce quantitatively how a particular physical property depends on another physical property, we have always first to derive an equation that shows the relationship between the two properties.

To deduce how density (ρ) would depend on pressure (p) we must therefore derive an equation connecting ρ and p. For this purpose we must start with equations for the two quantities $(\rho$ and $p)$ and then interconnect them.

Step 2 Select the key equation

We start with the defining equation for density which is

$$\rho = \frac{m}{V} \tag{1}$$

Step 3 Derive the required equation

Equation (1) shows the relationship between ρ and V. We want, however, the relationship between ρ and p. So the problem now is: how can we introduce p into equation (1)? To do this we must evidently know the relationship between V and p which by the ideal gas equation is

$$V = \frac{nRT}{p} \tag{2} \textbf{ Equation 5.1}$$

On eliminating V in equation (1) by substituting equation (2) we get

$$\rho = \frac{mp}{nRT} \tag{3}$$

Equation (3) gives the required relationship between ρ and p. Since $(m/n) = $ the molar mass M, we may simplify equation (3) and rewrite it as

$$\rho = \frac{Mp}{RT} \tag{4}$$

Equation (4) shows that the density of an ideal gas, at constant temperature, is *directly proportional* to the pressure.

Step 4

There is nothing to calculate.

Step 5 Review and check the solution

Exercise 5.3

By careful high vacuum techniques it is possible to reduce the pressure of a gaseous system to a value as low as 10^{-8} Pa. Estimate the number of molecules present in 1.0 dm^3 of such a system at 298 K.

Exercise 5.4

A meteorological balloon, filled with helium at 0 °C and at a pressure 1.0×10^5 Pa, is released into the atmosphere. It rises until its volume is doubled. If the temperature at this altitude is -5 °C, calculate the pressure.

Exercise 5.5

5.00 g of a gas is present in a closed 500 cm^3 vessel at 298 K. What further experimental data, if any, would you need to calculate (a) the pressure (b) the density, of the gas?

Exercise 5.6

An ideal gas has a pressure 2.00×10^5 Pa at 300 K. What is its concentration?

5.2 Molar masses of gases

The molar mass of a gas can be determined by the application of the ideal gas equation. Molar mass (M) is the mass (m) per mole. That is:

$$M = \frac{m}{n}$$ **Equation 3.3**

On substituting for n in the above equation by the ideal gas equation ($n = pV/RT$) we obtain

$$M = \frac{mRT}{pV}$$

All methods for the determination of molar masses of gases are fundamentally similar in that they are all based on the above equation. The four physical quantities m, p, V, and T must be measured to determine M.

Molar masses determined using experimental results at atmospheric pressure are only approximately correct. This is because all gases show deviation from the ideal gas equation at atmospheric pressure.

To determine *accurate* molar masses use is made of the fact that, for any gas, deviations from the ideal gas equation become less and less as the pressure is reduced. When the pressure approaches zero all gases behave ideally. Accurate molar masses could therefore be determined if the data required for the calculation could be obtained when the pressure of gas approaches zero. This can be done by a suitable extrapolation procedure, as is illustrated in Example 5.6.

Example 5.5 Determination of molar mass of a gas

1.00 dm^3 of a gas at 25 °C and at a pressure 1.00 atm has a mass 2.12 g. Calculate the molar mass of the gas (1 atm $= 1.013 \times 10^5$ Pa).

Solution

Step 1 Clarify and define the problem
We want the molar mass M of the gas.

Step 2 Select the key equation
The required quantity M is related to one of the items of data given (mass of gas, m) by the defining equation for M which is

$$M = \frac{m}{n}$$ (1) **Equation 3.3**

Step 3 Derive the equation for the calculation
Equation (1) cannot be used for the calculation of M since n is unknown. We must

therefore try to replace n in this equation by quantities given in the data. From the ideal gas equation it follows that

$$n = \frac{pV}{RT} \qquad \text{(2) Equation 5.1}$$

Substitution for n in equation (1) by equation (2) gives

$$M = \frac{mRT}{pV} \qquad (3)$$

Equation (3) could be used for calculating M since all the quantities required for the calculation are known.

Step 4 Collect the data, check the units, and calculate

$$m = 2.12 \times 10^{-3}\,\text{kg}$$

$$R = 8.314\,\text{kg m}^2\,\text{s}^{-2}\,\text{K}^{-1}\,\text{mol}^{-1}$$

$$T = 298\,\text{K}$$

$$p = 1.013 \times 10^5\,\text{kg m}^{-1}\,\text{s}^{-2}$$

$$V = 1.00 \times 10^{-3}\,\text{m}^3$$

$$M = \frac{mRT}{pV}$$

$$= \frac{(2.12 \times 10^{-3}\,\text{kg})\,(8.314\,\text{kg m}^2\,\text{s}^{-2}\,\text{K}^{-1}\,\text{mol}^{-1})\,(298\,\text{K})}{(1.013 \times 10^5\,\text{kg m}^{-1}\,\text{s}^{-2})\,(1.00 \times 10^{-3}\,\text{m}^3)}$$

$$= 5.19 \times 10^{-2}\,\text{kg mol}^{-1} = 51.9\,\text{g mol}^{-1}$$

Step 5 Review and check the solution

The units for M are correct and the value is reasonable.

Example 5.6 Accurate molar mass by extrapolation to zero pressure

At 273.15 K the density (ρ) of a gas is found to vary with pressure (p) in the following manner

$p/10^5$ Pa	0.800	0.600	0.400	0.200
$\rho/\text{g dm}^{-3}$	2.205	1.636	1.079	0.534

Calculate accurately, by a graphical extrapolation procedure, the molar mass of the gas.

Solution

Step 1 Clarify and define the problem

We have to calculate the molar mass of a gas accurately. To do so an extrapolation procedure is necessary. The necessity for graphical extrapolation to obtain accurate molar masses is first explained.

All gases show some deviation from ideal behaviour at atmospheric pressures. That is, they do not then obey rigorously the ideal gas equation. Molar masses calculated using the ideal gas equation with experimental results at atmospheric pressure (as in Example 5.5) are therefore only approximately correct.

A gas would obey the ideal gas equation accurately when its pressure approaches zero; the molar mass calculated by this equation using data corresponding to zero pressure would therefore be accurate. To obtain data corresponding to zero pressure we have to make use of an extrapolation procedure, as is illustrated in this example. We have to calculate the molar mass (M) from density (ρ) data at known pressures (p). To be able to do so, we have to derive the relationship between M, ρ, and p.

Step 2 Select the key equation

We start with the defining equation for the required quantity—molar mass M. That is

$$M = \frac{m}{n} \qquad \text{(1) \textbf{Equation 3.3}}$$

Step 3 Derive the equation for calculation

We have to introduce the data given (ρ and p) into equation (1). By definition

$$\rho = \frac{m}{V} \qquad (2)$$

On combining equations (1) and (2) we obtain

$$M = \frac{\rho V}{n} \qquad (3)$$

V is not given in the data but p is given. We therefore substitute (nRT/p) for V (by ideal gas equation) in equation (3); we then have

$$M = \frac{\rho RT}{p} \qquad (4)$$

By making use of this equation M could be calculated since all the quantities on the right side of the equation are given in the data.

Step 4 Collect the data, check the units, and calculate

Density data are given at four different pressures. At each pressure we could easily calculate M by making use of equation (4); the results then obtained are tabulated below:

$p/10^5$ Pa	0.800	0.600	0.400	0.200
$M/\text{g mol}^{-1}$	61.78	61.13	60.47	59.85

The value of M when $p \to 0$, which is the accurate value for M, could be obtained by plotting M vs p and extrapolating to zero pressure. If you make such a plot, you will find that the value of M when $p \to 0$, which is the accurate value for the molar mass, is 59.17 g mol^{-1}.

Step 5   Review and check the solution

The units for M are correct and the value is reasonable.

Exercise 5.7

The density of a gas is 1.97 g dm^{-3} at 273 K and at 1.01 × 10^5 Pa. Calculate the molar mass of the gas.

Exercise 5.8

1.00 dm^3 of a gas at 273 K and 1.00 × 10^5 Pa has a mass of 0.718 g. Which one of the following gases (a) CO_2 (b) CO (c) O_2 (d) CH_4 is the gas likely to be?

Exercise 5.9

Victor Meyer's method was often used as a laboratory experiment for determining approximately the molar mass of organic liquids which vaporize easily without decomposition.

0.0500 g of an organic liquid is found to displace, in a Victor Meyer's apparatus, 22.0 cm^3 of air measured at 25 °C and at a pressure of 750 mmHg. Calculate the molar mass of the liquid (1 mmHg = 1.33 × 10^2 Pa).

5.3   Dalton's law of partial pressures

Dalton's law of partial pressures states that the total pressure of a mixture of gases is equal to the sum of the *partial pressures* of the constituent gases.

The partial pressure of a constituent gas in a mixture is defined as the pressure that the gas would exert if it alone occupied a volume equal to that of the mixture. The partial pressure p_i of a gas i in a mixture having volume V and at temperature T is therefore given by the ideal gas equation as

$$p_i = n_i \frac{RT}{V} \qquad \text{Equation 5.1}$$

where n_i is the moles of gas i in the mixture.

From Dalton's law it follows that the total pressure p of a mixture of gases is given by

$$p = \Sigma \, p_i \qquad \text{Equation 5.2*}$$

Example 5.7   Application of Dalton's law

A 5.00 dm^3 vessel at 300 K contains 0.100 mol nitrogen molecules, 0.0500 mol oxygen molecules, and 0.0100 mol carbon dioxide molecules. Calculate (a) the partial pressure of each gas in the mixture (b) the total pressure.

Solution to Part (a)

Step 1   Clarify and define the problem

We want the partial pressure p_i of a gas i in a mixture of gases.

---

* The symbol Σ is used to represent a *sum*. Thus if three gases 1, 2, and 3, having partial pressures p_1, p_2, and p_3 are present in a mixture, the total pressure p would, by **Equation 5.2**, be given as

$$p = p_1 + p_2 + p_3$$

Step 2   Select the key equation
The partial pressure p_i of a gas i in a gas mixture is given by the ideal gas equation

$$p_i = \frac{n_i RT}{V}$$
Equation 5.1

Step 3   The equation for calculation
By the application of the **Equation 5.1** we could calculate the required partial pressures since all the quantities required for the calculation are known.

Step 4   Collect the data, check the units, and calculate

$$n_{N_2} = 0.100 \text{ mol}$$

$$n_{O_2} = 0.0500 \text{ mol}$$

$$n_{CO_2} = 0.0100 \text{ mol}$$

$$R = 8.314 \text{ J K}^{-1} \text{ mol}^{-1} = 8.314 \text{ kg m}^2 \text{ s}^{-2} \text{ K}^{-1} \text{ mol}^{-1}$$

$$T = 300 \text{ K}$$

$$V = 5.00 \text{ dm}^3 = 5.00 \times 10^{-3} \text{ m}^3$$

$$p_{N_2} = \frac{(0.100 \text{ mol}) (8.314 \text{ kg m}^2 \text{ s}^{-2} \text{ K}^{-1} \text{ mol}^{-1}) (300 \text{ K})}{(5.00 \times 10^{-3} \text{ m}^3)}$$

$$= 4.99 \times 10^4 \text{ Pa}$$

$$p_{O_2} = \frac{(0.0500 \text{ mol}) (8.314 \text{ kg m}^2 \text{ s}^{-2} \text{ K}^{-1} \text{ mol}^{-1}) (300 \text{ K})}{(5.00 \times 10^{-3} \text{ m}^3)}$$

$$= 2.49 \times 10^4 \text{ Pa}$$

$$p_{CO_2} = \frac{(0.0100 \text{ mol}) (8.314 \text{ kg m}^2 \text{ s}^{-2} \text{ K}^{-1} \text{ mol}^{-1}) (300 \text{ K})}{(5.00 \times 10^{-3} \text{ m}^3)}$$

$$= 0.499 \times 10^4 \text{ Pa}$$

Solution to Part (b)
The required quantity, the total pressure p, is easily calculated by the application of
Dalton's law, **Equation 5.2**

$$p = p_{N_2} + p_{O_2} + p_{CO_2}$$
Equation 5.2

$$= (4.99 \times 10^4 \text{ Pa}) + (2.49 \times 10^4 \text{ Pa}) + (0.499 \times 10^4 \text{ Pa})$$

$$= 7.98 \times 10^4 \text{ Pa}$$

Step 5   Review and check the solutions
The units are correct and the values appear reasonable.

Example 5.8 Relationship between partial pressure, mole fraction, and total pressure

Calculate the partial pressure of nitrogen in a mixture of gases if its mole fraction is 0.800 and the total pressure is 2.00×10^5 Pa.

Solution

Step 1 Clarify and define the problem

We want the partial pressure of nitrogen (p_{N_2}) in a mixture of gases.

Step 2 Select the key equation

To see how p_{N_2} could be calculated, we start with the defining equation for this quantity.

$$p_{N_2} = n_{N_2} \frac{RT}{V}$$
(1) **Equation 5.1**

n_{N_2} is the moles of nitrogen.

Step 3 Derive the equation for the calculation

Look at the data; none of the quantities required for the calculation of p_{N_2} by equation (1) are given there. The data give only the mole fraction of nitrogen (x_{N_2}) and the total pressure (p).

To calculate p_{N_2} from the data given, we should evidently have an equation which shows the relationship between p_{N_2} and the data given (x_{N_2} and p). We therefore try to replace the physical quantities in equation (1) with the quantities given in the data.

Consider first n_{N_2}, given in equation (1). It is related to one of the data (x_{N_2}) by the equation

$$n_{N_2} = x_{N_2} \times n$$
(2) **Equation 4.4**

where n is the total moles. On substituting equation (2) in (1) we get

$$p_{N_2} = x_{N_2} \times \frac{nRT}{V}$$
(3)

Can we now calculate p_{N_2} using equation (3)? No, because the term nRT/V is not given in the data. Can we relate this term to the data? Yes, because, by the ideal gas equation,

$$\frac{nRT}{V} = p$$
(4) **Equation 5.1**

where p, the total pressure, is given in the data. On substituting equation (4) in (3) we get

$$p_{N_2} = x_{N_2} \times p$$
(5)

Equation (5) is seen to be the desired equation because it shows how p_{N_2} is related to the data given; we can therefore use this equation for calculating p_{N_2}.

Step 4 Collect the data, check the units, and calculate
$$x_{N_2} = 0.800$$
$$p = 2.00 \times 10^5 \text{ Pa}$$
$$p_{N_2} = x_{N_2} \times p$$
$$= (0.800)(2.00 \times 10^5 \text{ Pa})$$
$$= 1.60 \times 10^5 \text{ Pa}$$

Step 5 Review and check the solution
The units are correct and the value is reasonable.

Note
This problem has been worked out in detail here in order to illustrate the five-step procedure. Most students will be able to answer this problem without writing down all the detail shown here. You only need to work through the five-step procedure in such a deliberate way when you are 'stuck', don't know what to do, or get a wrong answer.

Exercise 5.10

A 200.0 cm³ vessel is filled with a mixture of nitrogen and oxygen at 298.0 K and 1.013 × 10⁵ Pa. If the mass of the mixture is 0.2400 g, calculate (a) the mole fraction (b) the partial pressure (c) the percentage by mass (d) the percentage by volume, of oxygen in the vessel.

Exercise 5.11

A 1.00 dm³ flask containing hydrogen at 300 K and 1.20 × 10⁵ Pa is connected to a 5.00 dm³ flask containing nitrogen at 300 K and 1.50 × 10⁵ Pa. Calculate the final pressure if

(a) there is no reaction
(b) there is a quantitative reaction between nitrogen and hydrogen
(c) there is partial reaction and the partial pressure of hydrogen in the system is 8.00 × 10³ Pa
(d) there is partial reaction and the mole fraction of nitrogen in the system is 0.898.

5.4 Equation for pressure from the kinetic theory

The equation $pV = nRT$ was established by experiment. The pressure of an ideal gas can also be *calculated* by making use of the kinetic theory of gases. The equation then obtained, for a single gas at constant temperature, is

$$p = \frac{Nmc_{r.m.s.}^2}{3V} \qquad \textbf{Equation 5.3}$$

N is the total number of molecules in the vessel, m is the mass of each molecule, and $c_{r.m.s.}$ is the root mean square velocity of the gas molecules.

Note that **Equation 5.3** *interprets* the pressure of a gas in terms of the number N, mass m, and speed $c_{r.m.s.}$ of the constituent molecules; it is a *theoretical* equation in contrast to $pV = nRT$ which is an experimental equation.

Example 5.9 Calculation of root mean square velocity

Calculate the root mean square velocity of hydrogen molecules at 298 K. Molar mass of hydrogen molecule = 2.016 g mol⁻¹.

Solution

Step 1 Clarify and define the problem

We have to calculate the root mean square velocity $c_{r.m.s.}$ of hydrogen molecules.

Step 2 Select the key equation

To calculate $c_{r.m.s.}$ we have to start with an equation that involves this quantity.

$$c^2_{r.m.s.} = \frac{3pV}{Nm} \qquad \textbf{Equation 5.3}$$

$$\therefore \qquad c_{r.m.s.} = \left(\frac{3pV}{Nm}\right)^{1/2} \qquad (1)$$

Step 3 Derive the equation for the calculation

None of the quantities required for the calculation of $c_{r.m.s.}$ by equation (1) are given in the data; but T is given. So we introduce T into the equation by substituting nRT for pV (by the ideal gas **Equation 5.1**) Equation (1) then becomes

$$c_{r.m.s.} = \left[\frac{3nRT}{Nm}\right]^{1/2} \qquad (2)$$

There are three unknowns (n, N, and m) in the equation. The other quantity given in the data is the molar mass M. Can M be related to the unknowns and therefore be introduced into equation (2)? N is the number of molecules and m is the mass of one molecule and therefore

$$Nm = \text{mass of all the molecules}$$

$$= \text{mass of the gas}$$

From **Equation 3.3** it then follows that

$$\frac{\text{mass}}{n} = M \qquad (3) \textbf{ Equation 3.3}$$

Now on combining equations (2) and (3) we obtain

$$c_{r.m.s.} = \left(\frac{3RT}{M}\right)^{1/2} \qquad (4)$$

Equation (4) could be used to calculate $c_{r.m.s.}$ of hydrogen molecules since all the quantities on the right side of the equation are known.

Step 4 Collect the data, check the units, and calculate

$$M = 2.016 \text{ g mol}^{-1}$$

$$T = 298 \text{ K}$$

$$R = 8.314 \text{ kg m}^2 \text{ s}^{-2} \text{ K}^{-1} \text{ mol}^{-1}$$

$$c_{\text{r.m.s.}} = \left\{ \frac{3(8.314 \text{ kg m}^2 \text{ s}^{-2} \text{ K}^{-1} \text{ mol}^{-1}) (298 \text{ K})}{2.016 \times 10^{-3} \text{ kg mol}^{-1}} \right\}^{1/2}$$

$$= 1920 \text{ m s}^{-1}$$

Step 5 Review and check the solution

The units are correct.

The result shows that the molecules in a gas move about at very fast speeds. Since there are millions and millions of molecules present even in a very small sample of gas, it could be visualized that the gaseous state is a chaotic one from a molecular point of view. Many physical properties of gases (e.g. ready diffusion, thermal conductivity) can be understood in terms of the chaotic motion of the molecules present.

Exercise 5.12

A 1.00 dm^3 vessel contains 0.100 mole of a gas at a known pressure. What further data are necessary for the calculation of the root mean square velocity of the molecules?

Exercise 5.13

Would the root mean square velocity of the molecules of an ideal gas depend on (a) the temperature (b) the pressure (c) the amount of gas (d) the mass of the molecule?

Exercise 5.14

The root mean square velocity of hydrogen molecules, under a particular set of conditions, is 1.83×10^3 m s^{-1}. Calculate the root mean square speed of NH_3 molecules under the same conditions.

Exercise 5.15

A gas having a density 0.520 g dm^{-3} exerts a pressure of 0.242 atm. Calculate the root mean square velocity of the molecules. 1 atm $= 1.013 \times 10^5$ Pa.

Exercise 5.16

The percentage by volume of dry atmospheric air is 78.0% N_2, 21.0% O_2, 0.95% Ar, and 0.05% CO_2 (neglecting traces of H_2, Ne, etc.). Calculate

(a) the moles of each gas in 1 mol of air
(b) the percentage by mass of each gas
(c) the density of dry air at 25 °C.

Exercise 5.17

What would be the volume of a closed vessel if 10.0 g of ammonium chloride (NH_4Cl) heated to 500 °C exert a pressure of 1.00×10^5 Pa? Assume that all the NH_4Cl is dissociated into the gases ammonia and hydrogen chloride.

Exercise 5.18

How many grams of zinc must be dissolved in dilute sulphuric acid to obtain 1.00 dm^3 of hydrogen gas at 1.00×10^5 Pa and 300 K?

Exercise 5.19

When a dilute solution of sulphuric acid is electrolysed the reaction taking place is $2H_2O \rightarrow 2H_2 + O_2$. How many grams of water must be electrolysed to obtain 1.00 dm^3 of oxygen at 0 °C and 1.00×10^5 Pa pressure?

Exercise 5.20

0.100 mole of SO_2Cl_2 molecules (gas) is heated in a closed 10.0 dm^3 vessel at 200 °C. The degree (or fraction) of dissociation ($SO_2Cl_2(g) \rightleftharpoons SO_2(g) + Cl_2(g)$) is then 0.350. Calculate (a) the mole fraction (b) the partial pressure, of each gas in the vessel.

Exercise 5.21

15 cm^3 of a gaseous hydrocarbon was mixed with 100 cm^3 of oxygen. After combustion there was 45 cm^3 of carbon dioxide and 25 cm^3 of oxygen. Deduce the formula of the hydrocarbon. All volumes were measured under the same conditions.

Exercise 5.22

40.0 cm^3 of a gaseous mixture of carbon monoxide and ethyne (C_2H_2) was mixed with 100 cm^3 of oxygen and burnt completely. The final volume was 105 cm^3. How many cm^3 of carbon monoxide was present in the original mixture? All volumes were measured under the same conditions.

Exercise 5.23

A gas AO_2 dissociates on heating according to the equation $4AO_2(g) \rightleftharpoons 2A_2O_3(g) + O_2(g)$. Calculate the mole fraction of each gas in the mixture if AO_2 is 20.0% dissociated.

Exercise 5.24

Derive the relationship between the density of a gas and (a) its concentration (b) the root mean square velocity of the molecules in the gas.

Exercise 5.25

An important method for separating mixtures of gases is to make use of *Graham's law of effusion** (1846) according to which the rate of effusion of a gas is inversely proportional to the square root of its density. This method has been applied to separate the uranium isotope ^{235}U (which undergoes fission) from naturally occurring uranium which contains 99% ^{238}U (^{238}U does not undergo fission so readily).

Compare the relative rates of effusion, under the same conditions, of $^{235}UF_6$ and $^{238}UF_6$.

* Formerly known as Graham's law of diffusion.

Review of Chapter 5

Three equations were introduced in this chapter:

$$pV = nRT \qquad \textbf{5.1}$$

$$p = \Sigma \, p_i \qquad \textbf{5.2}$$

$$p = \frac{Nmc_{\text{r.m.s.}}^2}{3V} \qquad \textbf{5.3}$$

where p is the pressure, Pa
  V is the volume, m^3
  n is the amount of substance, mol
  R is the gas constant, J K^{-1} mol^{-1}
  T is the temperature, K
  p_i is the partial pressure of gas i, Pa
  N is the number of particles
  $c_{\text{r.m.s.}}$ is the root mean square velocity of the particles

You should understand clearly, remember, and be able to apply these equations fluently. Recognize that a thorough understanding of just these three simple equations provides an excellent introduction to the understanding of the physical behaviour of gases (strictly, of ideal gases).

All the examples and exercises in this chapter are based on just these three equations. Remember, however, that to solve certain problems the fundamental concepts and equations given in earlier chapters may have to be combined with the equations introduced in this chapter.

In some problems involving gases, chemical changes (dissociation, association, and other chemical reactions) are also involved. These influence the physical behaviour of a gas merely by altering the number of particles in the system. For calculating the moles in a system after reaction it is necessary to make use of the information provided by balanced equations (see section 4.1).

6 | Solutions

6.1 Raoult's law of vapour pressure

An important law of solutions is Raoult's law according to which the vapour pressure p_A due to a component A in a solution is directly proportional to the mole fraction x_A of that component in the solution.

This statement can be given in equation form as (see Example 2.4)

$$p_A = kx_A$$

where k is a constant.

Analysis of the equation (see Example 2.4) shows that $k = p_A^0$, where p_A^0 is the vapour pressure of the pure component A. Raoult's law may therefore be given as

$$p_A = p_A^0 x_A \qquad \textbf{Equation 6.1}$$

As a general rule, only very few solutions obey Raoult's law over the whole composition range; such solutions are known as ideal solutions.

Example 6.1 An equation derived from Raoult's law

Equation 6.1 (Raoult's law) relates p_A to the mole fraction of that component (x_A) in the solution. Starting with this equation derive, for a binary solution of A and B, the relationship between p_A and x_B, where x_B is the mole fraction of B in the solution.

Solution

Step 1 Clarify and define the problem

The problem is to replace x_A in the following equation with a term involving x_B.

$$p_A = p_A^0 x_A \qquad \text{(1) } \textbf{Equation 6.1}$$

Step 2 Select the key equation

To replace x_A in equation (1) by a term involving x_B we must derive an equation relating x_A and x_B. To do so, we start with the defining equations for these two quantities which are

$$x_A = \frac{n_A}{n_A + n_B} \qquad \textbf{Equation 4.4}$$

$$x_B = \frac{n_B}{n_A + n_B} \qquad \textbf{Equation 4.4}$$

Step 3 Derive the required equation

From these two equations for x_A and x_B it can be seen that

$$x_A + x_B = 1$$

∴
$$x_A = 1 - x_B \qquad (2)$$

On substituting for x_A in equation (1) by equation (2) we obtain

$$p_A = p_A^0(1 - x_B) \qquad (3)$$

(3) is the required equation since it shows the relation between p_A and x_B.

Step 4

Does not apply.

Step 5 Review and check the solution

Note

Suppose that a solution of a *non-volatile* solute B is considered. The vapour pressure of the solution would then be due only to the solvent. Equation (3) is then generally written as

$$p = p^0(1 - x_B)$$

where p^0 and p are the vapour pressures of the pure solvent and the solution respectively.

Example 6.2 Calculation of mole fraction in vapour from mole fraction in solution

Calculate the mole fraction of benzene in the vapour phase above an equimolar solution of benzene and methylbenzene (toluene) at 298 K. At this temperature the vapour pressures of pure benzene and pure toluene are 96.0 Torr and 31.0 Torr respectively (1 Torr = pressure exerted by a column of mercury 1 mm high = 1.33 × 10^2 Pa).*

Solution

Step 1 Clarify and define the problem

There will be benzene and toluene in the vapour—define the mole fractions in the vapour as x_B and x_T respectively. There will also be benzene and toluene in solution—define the mole fractions in solution as x_B(solution) and x_T(solution) respectively.

The problem is presented in Fig. 6.1. Recognize that the word 'equimolar' provides the information that x_B(solution) = x_T(solution). Since it follows from the definition of mole fraction that $x_B + x_T = 1$, it is clear that

$$x_B(\text{solution}) = x_T(\text{solution}) = 0.500$$

* Torr is not an SI unit, but it is used in many books and is introduced here for practice.

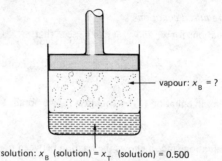

solution: x_B (solution) = x_T (solution) = 0.500

Fig. 6.1   A visual presentation of the problem in Example 6.2

Step 2   Select the key equation

The mole fraction of benzene (x_B) in the *vapour* phase is required; we should therefore focus our attention on the vapour phase. The defining equation for the required quantity—x_B in the vapour phase—is

$$x_B = \frac{n_B}{n_B + n_T} \qquad \text{(1) Equation 4.4}$$

n_B and n_T respectively are the moles of benzene and toluene present in a certain volume of the vapour.

Step 3   Derive the equation for the calculation

To calculate x_B by equation (1) we require n_B and n_T; these quantities, however, are not given in the data. So we must try to relate them to the data given; for this purpose we apply the ideal gas equation separately to each component in the gas mixture. We then obtain

$$n_B = \frac{p_B V}{RT} \qquad \text{(2) Equation 5.1}$$

and

$$n_T = \frac{p_T V}{RT} \qquad \text{(3) Equation 5.1}$$

p_B and p_T are the partial pressures of benzene and toluene respectively.

If we substitute equations (2) and (3) in equation (1) and simplify we obtain

$$x_B = \frac{p_B}{p_B + p_T} \qquad \text{(4)}$$

p_B and p_T are related to the data given (x_B(solution), x_T(solution), p_B^0, p_T^0) by **Equation 6.1**; the application of this equation shows that

$$p_B = p_B^0 x_B(\text{solution}) \qquad \text{(5) Equation 6.1}$$

$$p_T = p_T^0 x_T(\text{solution}) \qquad \text{(6) Equation 6.1}$$

p_B^0 and p_T^0 are the vapour pressures of pure benzene and pure toluene respectively.

On substituting for p_B and p_T in equation (4) by equations (5) and (6) we obtain

$$x_B = \frac{p_B^0 \, x_B(\text{solution})}{p_B^0 \, x_B(\text{solution}) + p_T^0 \, x_T(\text{solution})} \qquad (7)$$

Equation (7) could be used to calculate the required quantity (x_B) since all the quantities on the right side of the equation are known.

Step 4 Collect the data, check the units, and calculate

From the given data it is seen that

$$p_B^0 = 96.0 \text{ Torr}$$

$$p_T^0 = 31.0 \text{ Torr}$$

$$x_B(\text{solution}) = x_T(\text{solution}) = 0.500$$

Since the required quantity (x_B) has no units (we know this because x is defined as a ratio), we can give the values for the physical quantities in any units. The p^0 values can, therefore, be given in Torr units; they need not be converted into SI units.

On inserting the values given above in equation (7) and on calculating we obtain

$$x_B = \frac{(96.0 \text{ Torr})(0.500)}{(96.0 \text{ Torr})(0.500) + (31.0 \text{ Torr})(0.500)}$$

$$= \frac{48.0 \text{ Torr}}{(48.0 + 15.5) \text{ Torr}}$$

$$= 0.756$$

Step 5 Review and check the solution

The answer is reasonable because by definition x_B must lie between 0 and 1, and also has no units. The number of significant figures in the answer is also appropriate.

Once the problem is clearly presented as in Fig. 6.1, the solution becomes easier. If not, there may be confusion because there are two phases and therefore two values for x_B—one in the solution phase and one in the vapour phase. The figure emphasizes that we must focus our attention on the vapour phase and start deducing the solution from there.

Exercise 6.1

Which, in each case, has the higher vapour pressure:

(a) pure water or an aqueous salt solution?
(b) a 5% (by mass) solution of glucose $(C_6H_{12}O_6)$ or a 5% (by mass) solution of sucrose $(C_{12}H_{22}O_{11})$?

Exercise 6.2

A mixture of chlorobenzene and bromobenzene behaves ideally. A solution contains 50.0 grams of chlorobenzene (C_6H_5Cl) and 50.0 grams of bromobenzene (C_6H_5Br). Calculate the partial vapour pressures and the total vapour pressure. What is the mole fraction of chlorobenzene in the vapour phase? The vapour pressures of pure chlorobenzene and bromobenzene are 3.20×10^4 Pa and 2.40×10^4 Pa respectively at the temperature of the experiment.

Exercise 6.3

A binary solution of A in B behaves ideally (i.e. obeys Raoult's law). Show graphically the variation with x_A of

(a) the partial vapour pressure of component A
(b) the partial vapour pressure of component B
(c) the total vapour pressure of the solution.

Remember that to draw a graph between any two variables, it is necessary first to derive the equation showing the relation between the variables.

Exercise 6.4

A solute B is dissolved in a solvent A. Derive the relationship between the mole fraction of the solute (x_B) in the solution and the molar mass of the solute (M_B). What data are needed to calculate M_B from x_B?

Exercise 6.5

An application of vapour pressure measurements is in the determination of the molar mass of a non-volatile solute in a solution. The data required and the method of calculation will become clear once this exercise is solved.

A solution containing 5.00 g of a non-volatile solute in 100.0 g of water has a vapour pressure 3.145×10^4 Pa. The vapour pressure of pure water at the same temperature is 3.160×10^4 Pa. Calculate the molar mass of the solute.

Exercise 6.6

When a liquid is heated, the vapour pressure increases with temperature. At the boiling point the vapour pressure is equal to the external pressure.

(a) Which would have the higher boiling point—an aqueous sugar solution, or water?
(b) Why is the boiling point of water at the top of a mountain lower than that at sea level?

6.2   Freezing point and boiling point of solutions

The freezing point of a solvent is depressed when a non-volatile solute is added to it. The depression of freezing point is found experimentally to be directly proportional, in dilute solutions, to the molality of the non-volatile solute in the solution. It does not depend on the nature of the solute.

The statement given above could be represented in equation form as follows. Let T_f^0 and T_f be the freezing points of the pure solvent and a solution of molality m respectively. $(T_f^0 - T_f)$ is then the depression of freezing point and since this is proportional to the molality m of the solution we may write

$$T_f^0 - T_f = K_f m \qquad\qquad \textbf{Equation 6.2}$$

K_f, the constant of proportionality, is known as the molal freezing point depression constant.

The boiling point of a solvent is elevated when a non-volatile solute is added to it. The elevation of boiling point $(T_b - T_b^0)$ is also found experimentally to be directly proportional to the molality. That is

$$T_b - T_b^0 = K_b m \qquad \textbf{Equation 6.3}$$

where T_b^0 and T_b are the boiling points of the pure solvent and of the solution respectively. K_b is known as the molal boiling point elevation constant.

Equations 6.2 and 6.3 are applicable accurately only for very dilute solutions.*

Freezing point and boiling point measurements of solutions provide one method for the determination of the molar mass of a non-volatile solute in solution. The data required and the method of calculation are illustrated in Example 6.3.

Example 6.3 Molar mass of a solute from boiling point elevation

The boiling point of an aqueous solution containing 5.00 g of a non-volatile solute in 100.0 g of water is 373.47 K. The boiling point of water is 373.15 K. Calculate the molar mass of the solute. For water, $K_b = 0.51$ K kg mol^{-1}.

Solution

Step 1 *Clarify and define the problem*

We have to calculate the molar mass M of the solute from boiling point data. To do so we have to derive an equation which relates M to boiling point data.

Step 2 *Select the key equation*

As recommended in Chapter 1, we start with the defining equation for the required quantity—molar mass M.

$$M = \frac{m_{\text{solute}}}{n_{\text{solute}}} \qquad \text{(1) Equation 3.3}$$

Step 3 *Derive the equation for the calculation*

Look at equation (1) and the data given. m_{solute} is given in the data; n_{solute} is not. We should try to relate n_{solute} to the data given.

n_{solute} could be related to the boiling point data given through the molality m since both n_{solute} and the boiling point data are related to the molality m.

$$n_{\text{solute}} = m \times m_{\text{solvent}} \qquad \text{(2) Equation 4.3}$$

$$m = \frac{T_b - T_b^0}{K_b} \qquad \text{(3) Equation 6.3}$$

On substituting for m in equation (2) by equation (3) we obtain

$$n_{\text{solute}} = \frac{T_b - T_b^0}{K_b} \times m_{\text{solvent}} \qquad \text{(4)}$$

Combination of equations (1) and (4) would then give the required equation for calculation.

$$M = \frac{m_{\text{solute}} \times K_b}{(T_b - T_b^0) \times m_{\text{solvent}}} \qquad \text{(5)}$$

* In dilute solutions m is approximately the same as c and sometimes c is used in freezing point depression and boiling point elevation calculations.

Step 4 Collect the data, check the units, and calculate

$$m_{solute} = 5.00 \text{ g} = 5.00 \times 10^{-3} \text{ kg}$$

$$K_b = 0.51 \text{ K kg mol}^{-1}$$

$$T_b = 373.47 \text{ K}$$

$$T_b^0 = 373.15 \text{ K}$$

$$m_{solvent} = 100 \text{ g} = 100 \times 10^{-3} \text{ kg}$$

By the application of equation (5) we obtain

$$M = \frac{(5.00 \times 10^{-3} \text{ kg})(0.51 \text{ K kg mol}^{-1})}{(373.47 \text{ K} - 373.15 \text{ K})(100.0 \times 10^{-3} \text{ kg})}$$

$$= 7.97 \times 10^{-2} \text{ kg mol}^{-1}$$

$$= 79.7 \text{ g mol}^{-1}$$

Step 5 Review and check the solution

Example 6.4 Calculations involving freezing

An industrial method for *concentrating* aqueous solutions (such as extracts of fruits and sugar cane) is by freezing. Pure solvent (ice) then separates out; the solution remaining would therefore become more concentrated. The reason for using this method is that the cane sugar would decompose if the water was distilled off by boiling.

1000 g of an aqueous solution of cane sugar contained 200 g of solute. When this solution was cooled and maintained at a constant temperature below its freezing point, 600 g of ice separated out. Calculate the molality of the remaining solution. Freezing point of water is 273.15 K, K_f is 1.86 K kg mol^{-1}, and the molar mass of cane sugar is 342 g mol^{-1}.

Solution

Step 1 Clarify and define the problem

In the problem given there are two states—let us call these the initial state and the final state. The data given corresponding to these two states are presented clearly by the following figure.

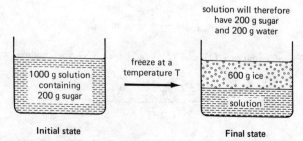

Fig. 6.2 A visual presentation of the problem in Example 6.4

From the *law of conservation of mass* (see section 4.2) it follows that

(a) mass of solute in the = mass of solute in the
    initial state   final state

(b) total mass in the = total mass in the
    initial state   final state.

By making use of these two relationships it is easy to see that the solution in the final state would have 200 grams of solute (sugar) and 200 grams of solvent (water).

Step 2 Select the key equation
We want the molality m of the solution in the final state. To calculate it we start with the defining equation for m which is

$$m = \frac{n_{solute}}{m_{solvent}} \qquad \text{(1) Equation 4.3}$$

Step 3 Derive the equation for calculation
In the solution in the final state the mass of solvent ($m_{solvent}$) was shown earlier to be 200 grams. To calculate m by equation (1) we therefore require n_{solute} which is the moles of solute associated with the 200 grams of solvent. By definition

$$n_{solute} = \frac{m_{solute}}{M} \qquad \text{(2) Equation 3.3}$$

On replacing n_{solute} in equation (1) by equation (2) we obtain

$$m = \frac{m_{solute}}{M m_{solvent}} \qquad \text{(3)}$$

Equation (3) can be used to calculate the required molality m.

Step 4 Collect the data, check the units, and calculate
In the solution in the final state

$$m_{solute} = 200 \text{ g}$$

$$m_{solvent} = 200 \text{ g}$$

Also from the given data

$$M = 342 \text{ g mol}^{-1}$$

$$\therefore \quad m = \frac{200 \text{ g}}{(342 \text{ g mol}^{-1})\, 200 \text{ g}}$$

$$= 2.92 \times 10^{-3} \text{ mol g}^{-1}$$

$$= 2.92 \text{ mol kg}^{-1}$$

The required molality of the solution in the final state is therefore 2.92 mol kg^{-1}.

Step 5 Review and check the solution

The units are correct.

A clear pictorial presentation (Fig. 6.2) of the problem in Step 1 helped to focus on the problem and make the solution easier. Neither the freezing point of water nor K_f given in the data were needed to solve this problem.

Exercise 6.7

An aqueous solution of sodium chloride has a freezing point -0.93 °C. The freezing point of water is 0.00 °C and $K_f = 1.86$ K kg mol^{-1}. Calculate the molality of the solution.

Exercise 6.8

Calculate the normal boiling point of sea water, given that it contains 3.5% by mass NaCl and 0.13% by mass $MgCl_2$. The normal boiling point of water is 100.00 °C and $K_b = 0.51$ K kg mol^{-1}.

Exercise 6.9

A salt solution exposed to the atmosphere is boiled. Explain why the boiling point would go on increasing.

Exercise 6.10

In many countries the temperatures during winter are often below the freezing point of water (0 °C). To prevent the water in the radiator of cars from freezing, compounds such as ethane-1,2-diol (ethylene glycol, $C_2H_4(OH)_2$) and propane-1,2,3-triol (glycerol, $C_3H_5(OH)_3$) (these compounds have high solubility and are non-corrosive) are added. Is a 30% (by weight) aqueous solution of ethylene glycol satisfactory for this purpose in Siberia where the temperature during winter is often -50 °C? $K_f = 1.86$ K kg mol^{-1}.

Exercise 6.11

100 grams of an aqueous solution is cooled and maintained at -3.52 °C. Calculate the molality of the solution after ice separates out. Freezing point of water is 0.00 °C and $K_f = 1.78$ K kg mol^{-1}. What further data would you need to calculate the mass of ice that would separate out?

Exercise 6.12

An aqueous solution of a monobasic acid (molar mass of acid, 60.0 g mol^{-1}) containing 1.500 grams dissolved in 100.0 grams of water has a freezing point of -0.68 °C. Calculate the degree of dissociation of the acid in the solution. T_f^0 0.00 °C and K_f 1.78 K kg mol^{-1}.

Exercise 6.13

A standard method for checking the purity of a solid is to determine its melting point (note: melting point and freezing point are identical). If impurities are present the melting point is depressed. A sample of naphthalene which contains an impurity has a melting point 78.6 °C.

Calculate the percentage purity if the molar mass of the impurity is 200 g mol⁻¹. K_f for naphthalene is 6.80 K kg mol⁻¹ and the melting point of naphthalene is 80.8 °C.

Exercise 6.14

The freezing point of an aqueous solution is −0.36 °C. Calculate (a) the normal boiling point of the solution (b) the vapour pressure at 25 °C. The vapour pressure of water at 25 °C is 3.310×10^3 Pa. K_f 1.86 K kg mol⁻¹ and K_b 0.52 K kg mol⁻¹. The molar masses of the solute and water are 180 g mol⁻¹ and 18.0 g mol⁻¹ respectively.

6.3 Osmotic pressure

Experiments show that the osmotic pressure (Π) of a solution is given by the equation

$$\Pi = RTc \qquad\qquad \textbf{Equation 6.4}$$

where c is the total concentration of all the solute species present, R is the gas constant, and T is the temperature.

Equation 6.4 is applicable accurately only in very dilute solutions. This equation is similar to **Equation 5.1**, $pV = nRT$ and can also be written as

$$\Pi = \frac{n}{V} \times RT$$

Example 6.5 Calculation of osmotic pressure

Calculate the osmotic pressure of a 0.100 mol dm⁻³ aqueous solution of sodium chloride at 300 K.

Solution

Step 1 Clarify and define the problem
We have to calculate the osmotic pressure Π of a solution.

Step 2 Select the key equation
The required osmotic pressure could be calculated by the application of the equation

$$\Pi = RTc \qquad\qquad \text{(1) } \textbf{Equation 6.4}$$

Here c is the *total* concentration of all the solute species in the solution.

Step 3 Derive the equation for the calculation
To calculate Π, we must first find out c. Since NaCl consists of Na⁺ and Cl⁻ ions it follows that in solution

$$c = c_{Na^+} + c_{Cl^-} \qquad\qquad (2)$$

From equations (1) and (2) it is clear that

$$\Pi = RT(c_{Na^+} + c_{Cl^-}) \qquad\qquad (3)$$

Step 4 Collect the data, check the units, and calculate

By the application of equation (3) we can calculate Π. If each quantity is given in SI unit Π obtained would be in SI unit.

$$T = 300 \text{ K}$$

$$R = 8.314 \text{ J K}^{-1} \text{mol}^{-1}$$

$$= 8.314 \text{ kg m}^2 \text{ s}^{-2} \text{ K}^{-1} \text{mol}^{-1} \qquad \text{(see note)}$$

$$c_{Na^+} = c_{Cl^-} = 0.100 \text{ mol dm}^{-3}$$

$$= 100 \text{ mol m}^{-3}$$

$$\therefore \qquad \Pi = (8.314 \text{ kg m}^2 \text{ s}^{-2} \text{ K}^{-1} \text{mol}^{-1}) \, (300 \text{ K}) \, (200 \text{ mol m}^{-3})$$

$$= 4.99 \times 10^5 \text{ kg m}^{-1} \text{s}^{-2}$$

$$= 4.99 \times 10^5 \text{ Pa} \qquad \text{(see note)}$$

Step 5 Review and check the solution

The units are correct and the number of significant figures is appropriate.

Note

If you find any difficulty in understanding the replacement of J by kg m^2 s^{-2} or of kg m^{-1} s^{-2} by Pa, study section 2.4 once again.

Exercise 6.15

For patients who cannot be fed by mouth, glucose solution is often slowly injected into the blood stream through a vein. Any solution so injected should have the same osmotic pressure as that of blood. What should be the molar concentration of the glucose solution if the osmotic pressure of blood is 7.77×10^5 Pa at body temperature (37 °C)?

Exercise 6.16

Osmotic pressure measurements are important because they provide an accurate method for determining the molar masses of substances—particularly of macromolecules such as rubber, plastics, proteins, enzymes, etc.

The osmotic pressure at 25 °C of a solution of natural rubber in benzene (5.92 g dm^{-3}) is 2.93×10^2 Pa. Calculate the molar mass of the sample of rubber.

Explain why you would expect the molar mass so calculated to be inaccurate. What further experimental data must be obtained if one wants to determine M accurately?

Review of Chapter 6

Four simple equations were introduced in Chapter 6. They are

$$p_A = p_A^0 x_A \qquad \qquad \textbf{6.1}$$

$$T_f^0 - T_f = K_f m \qquad \qquad \textbf{6.2}$$

$$T_b - T_b^0 = K_b m \qquad \qquad \textbf{6.3}$$

$$\Pi = RTc \qquad \qquad \textbf{6.4}$$

where p_A is the vapour pressure due to component A in solution, Pa

   p_A^0 is the vapour pressure of pure A, Pa

   x_A is the mole fraction of A

   T_f^0 is the freezing point of pure solvent, K

   T_f is the freezing point of solution, K

   K_f is the freezing point depression constant, K mol^{-1} kg

   T_b is the boiling point of pure solvent, K

   T_b^0 is the boiling point of solution, K

   K_b is the boiling point elevation constant, K mol^{-1} kg

   m is the molality of solution, mol kg^{-1}

   Π is the osmotic pressure, Pa

   R is the gas constant, J K^{-1} mol^{-1}

   T is the temperature, K

   c is the total concentration of all solute species, mol m^{-3}

You should understand and be able to recall these equations and apply them whenever necessary. A large amount of fundamental knowledge about solutions is summarized in these equations.

All examples and exercises in this chapter are based on these simple equations. Application of these equations is straightforward and easy.

The difficulty which students sometime have in working problems involving these equations is generally in the concentration term (x, m, or c)—how to calculate concentrations under the specified conditions. This once again emphasizes the necessity for a thorough knowledge of the various concentration units given in section 4.5.

7 | Physical and chemical equilibria

7.1 Law of chemical equilibrium

Many properties studied in chemistry are equilibrium properties. An equilibrium property is one whose value does not change with time. Common examples of equilibrium properties are: pressure, vapour pressure, equilibrium constant, dissociation constant, solubility product, etc.

The law of chemical equilibrium is one of the most important generalizations in chemistry. For a system in equilibrium represented in general by the equation

$$a\text{A} + b\text{B} + \ldots \rightleftharpoons \ldots y\text{Y} + z\text{Z}$$

the law of chemical equilibrium states that

$$\frac{\ldots \times [\text{Y}]^y \times [\text{Z}]^z}{[\text{A}]^a \times [\text{B}]^b \times \ldots} = K \qquad \textbf{Equation 7.1}$$

where K is a constant at a constant temperature. This constant is known as the equilibrium constant. (Square brackets [ ] denote concentration.)

Remember the following:

1 **Equation 7.1** is applicable to *all* types of equilibria.
2 The concentrations given in **Equation 7.1** are equilibrium concentrations and not initial concentrations. (Strictly the concentrations in the equation should be 'idealized' concentrations—i.e. concentrations after correcting for any non-ideal behaviour. In this book the error due to non-ideal behaviour will be ignored.)
3 The concentrations in **Equation 7.1** can be expressed in any suitable units, but most commonly either concentration (c) or partial pressure p (possible only for gases) is used. When expressed in terms of p, the equilibrium constant is denoted by K_p; and when expressed in terms of c it is denoted K_c.
4 If any concentration term in the equation remains constant during the reaction, it will have a constant effect on the equilibrium. It is the usual practice, therefore, to define the equilibrium constant only in terms of concentration terms that can change during the reaction. In other words, when writing the expression for K you should not include any concentration terms that remain constant. It can be shown that the concentration of a pure solid in any system, and the concentration of water in dilute aqueous solutions are constants. These concentration terms should not, therefore, appear in the expression for an equilibrium constant.

Exercise 7.1

Calculate the molar concentration of iron in pure iron. The density of iron is 7.86 g cm^{-3} and its molar mass is 55.8 g mol^{-1}. Recognize clearly that the molar concentration of iron (and also of all solids) is independent of the amount of iron and is a constant at constant temperature.

Exercise 7.2

Write the expression for the equilibrium constant for each of the following equilibria. For gaseous equilibria, express concentrations in terms of partial pressures; for other equilibria use concentration (c).

(a) $N_2O_4(g) \rightleftharpoons 2NO_2(g)$
(b) $NH_4Cl(s) \rightleftharpoons NH_3(g) + HCl(g)$
(c) $CaCO_3(s) \rightleftharpoons CaO(s) + CO_2(g)$
(d) $AgCl(s) \rightleftharpoons Ag^+(aq) + Cl^-(aq)$
(e) $CH_3COOH(aq) \rightleftharpoons CH_3COO^-(aq) + H^+(aq)$
(f) $H_2O(l) \rightleftharpoons H^+(aq) + OH^-(aq)$
(g) $Zn(s) + 2Ag^+(aq) \rightleftharpoons Zn^{++}(aq) + 2Ag(s)$
(h) $H_2O(l) \rightleftharpoons H_2O(g)$
(i) NH_3 (in ether) $\rightleftharpoons NH_3$ (in water)

7.2 Determination of equilibrium constant

Equation 7.1 shows that to determine the equilibrium constant of a reaction we must measure the concentrations at equilibrium of the various reactants and products. The required equilibrium concentrations can be determined either by chemical analysis of the equilibrium mixture (see Example 7.1) or by measuring any physical property (e.g. pressure, volume, density, colour...) that depends on the relative amounts of reactants and products (see Example 7.2).

Example 7.1 Determination of equilibrium constant by chemical analysis

Hydrogen iodide (n moles) was maintained at $400\,°C$ in a sealed glass bulb of volume V until equilibrium was attained. The equilibrium was 'frozen' by rapid cooling of the equilibrium mixture. This was then titrated with sodium thiosulphate solution and it was found that x moles of iodine were present at equilibrium. Show how the equilibrium constant K_c could be calculated if n and x are known.

Solution

Step 1 Clarify and define the problem
To show how K_c could be calculated from n and x data we have to derive an equation which shows how K_c is related to n and x.

Step 2 Select the key equation
We start with the defining equation for the required quantity. For the reaction

$$2HI \rightleftharpoons H_2 + I_2$$

the equilibrium constant K_c is given by the equation

$$K_c = \frac{c_{H_2} c_{I_2}}{c_{HI}^2} \qquad \text{(1) Equation 7.1}$$

where c_{H_2} refers to the concentration of hydrogen *at equiliorium* and likewise for c_{I_2} and c_{HI}.

Step 3 Derive the required equation

To calculate K_c by equation (1) we should therefore know c_{H_2}, c_{I_2}, and c_{HI} in an equilibrium mixture. Since x mol of I_2 are present in a volume V it follows by the application of the defining equation for molar concentration that

$$c_{I_2} = \frac{x}{V} \qquad \text{(2) Equation 4.2}$$

Expressions for c_{H_2} and c_{HI} in terms of x can also be obtained by making use of the balanced equation (see section 4.1) for the dissociation which shows that

$$c_{H_2} = \frac{x}{V} \tag{3}$$

$$c_{HI} = \frac{n - 2x}{V} \tag{4}$$

On substituting equations (2), (3), and (4) in equation (1) we obtain

$$K_c = \frac{c_{H_2} \times c_{I_2}}{c_{HI}^2} = \frac{\dfrac{x}{V} \times \dfrac{x}{V}}{\left(\dfrac{n - 2x}{V}\right)^2}$$

$$= \frac{x^2}{(n - 2x)^2} \tag{5}$$

K_c could therefore be calculated by equation (5) provided that the initial amount of hydrogen iodide (n) and the amount of iodine formed at equilibrium (x) are known.

Step 4

Does not apply.

Step 5 Review and check the solution

Note

The equation for the equilibrium constant (equation 5) does not include a volume or pressure term. The equilibrium composition would not, therefore, depend on volume or pressure. This conclusion would be found to be true for all equilibria where the total moles of reactants is equal to the total moles of products.

Example 7.2 Determination of equilibrium constant by a physical method

Dinitrogen tetroxide (N_2O_4) is a gas, above 21 °C, which dissociates into nitrogen dioxide according to the equation

$$N_2O_4(g) \rightleftharpoons 2NO_2(g)$$

Suppose a moles of N_2O_4 are maintained at a temperature T in a closed vessel of volume V. The pressure at equilibrium, which could be measured by a manometer attached to the vessel, is found to be p. Indicate how K_p for the dissociation could be calculated if values of a, T, V, and p are known.

Solution

Step 1   Clarify and define the problem

To indicate how K_p could be calculated, our objective should be to derive an equation which shows how K_p is related to the data (a, T, V, p).

Step 2   Select the key equation

We start with the defining equation for K_p which, for the reaction $N_2O_4 \rightleftharpoons 2NO_2$, is

$$K_p = \frac{p_{NO_2}^2}{p_{N_2O_4}} \qquad \text{(1) Equation 7.1}$$

Step 3   Derive the required equation

Our problem essentially is: how can we introduce the data a, T, V, and p into equation (1)? T and V could be introduced by making use of the equations

$$p_{NO_2} = \frac{n_{NO_2} RT}{V} \qquad \text{(2) Equation 5.1}$$

$$p_{N_2O_4} = \frac{n_{N_2O_4} RT}{V} \qquad \text{(3) Equation 5.1}$$

On replacing p_{NO_2} and $p_{N_2O_4}$ in equation (1) by equations (2) and (3), and on simplifying we obtain

$$K_p = \frac{n_{NO_2}^2 RT}{n_{N_2O_4} V} \qquad \text{(4)}$$

In equation (4), the only unknowns are n_{NO_2} and $n_{N_2O_4}$ (R is the gas constant and its value can be looked up). We can relate them to a (the moles of N_2O_4 that was present initially) by making use of the balanced equation for the dissociation

$$N_2O_4 \rightleftharpoons 2NO_2$$

Suppose that from the initial a moles, b moles of N_2O_4 had dissociated at equilibrium. It then follows from the balanced equation that

$$n_{NO_2} = 2b \qquad \text{(5)}$$

$$n_{N_2O_4} = (a - b) \qquad \text{(6)}$$

On substituting for n_{NO_2} and $n_{N_2O_4}$ in equation (4) by equations (5) and (6), and on simplifying we get

$$K_p = \frac{4b^2 RT}{(a - b) V} \qquad \text{(7)}$$

In equation (7) the only quantity not given in the data is b. Can we relate it to the data given? Yes, by the application of the ideal gas equation. Since the total number of moles present at equilibrium n is given by

$$n = n_{NO_2} + n_{N_2O_4} = 2b + (a - b) = (a + b)$$

it follows by the application of the ideal gas equation to the whole equilibrium mixture that

$$pV = (a + b)RT \qquad \text{Equation 5.1}$$

$$\therefore \qquad b = \frac{pV}{RT} - a \qquad (8)$$

On substituting for b in equation (7) by equation (8) one obtains

$$K_p = \frac{4RT\left(\dfrac{pV}{RT} - a\right)^2}{\left\{a - \left(\dfrac{pV}{RT} - a\right)\right\} \times V}$$

The above equation is the required equation since it indicates how K_p could be calculated from a, T, V, and p.

Step 4
Does not apply.

Step 5 Review and check the solution

Example 7.3 Calculation of an equilibrium constant

The Haber process is of great technological importance because it converts the relatively inexhaustible supply of nitrogen in the atmosphere into the important compound ammonia.

When a mixture containing 1.00 moles of nitrogen (N_2) molecules and 3.00 moles of hydrogen (H_2) molecules is heated in a vessel at 773 K and at a pressure 3.55×10^7 Pa, 30% of the nitrogen is converted, at equilibrium, into ammonia. Calculate (a) K_c (b) K_p, for the reaction.

Solution to Part (a)

Step 1 Clarify and define the problem
We have to calculate K_c, the equilibrium constant in terms of concentrations.

Step 2 Select the key equation
The defining equation for the required quantity K_c, for the equilibrium $N_2 + 3H_2 \rightleftharpoons 2NH_3$, is

$$K_c = \frac{c_{NH_3}^2}{c_{N_2}\, c_{H_2}^3} \qquad (1) \text{ Equation 7.1}$$

where c_{NH_3}, etc. represent concentrations *at equilibrium*.

Step 3 Derive the equation for the calculation

Equation (1) shows that to calculate K_c for the given reaction we require the following concentrations at equilibrium: c_{NH_3}, c_{N_2}, and c_{H_2}. How can these concentrations be calculated? To answer this question look at the defining equations for these concentration terms which are

$$c_{N_2} = \frac{n_{N_2}}{V} \qquad \text{(2) Equation 4.2}$$

$$c_{H_2} = \frac{n_{H_2}}{V} \qquad \text{(3) Equation 4.2}$$

$$c_{NH_3} = \frac{n_{NH_3}}{V} \qquad \text{(4) Equation 4.2}$$

These equations show that to calculate the required equilibrium concentrations we require the volume V of the vessel and the number of moles of N_2, H_2, and NH_3 at equilibrium. These quantities are not given in the data; they can, however, be calculated from the data given.

Consider first V. It is related to the data given (p, T) by the equation

$$V = \frac{nRT}{p} \qquad \text{(5) Equation 5.1}$$

where n is the total moles present.

Replacement of V in equations (2), (3), and (4) by equation (5) then gives

$$c_{N_2} = \frac{n_{N_2}\, p}{nRT} \qquad (6)$$

$$c_{H_2} = \frac{n_{H_2}\, p}{nRT} \qquad (7)$$

$$c_{NH_3} = \frac{n_{NH_3}\, p}{nRT} \qquad (8)$$

On replacing c_{N_2}, c_{H_2}, and c_{NH_3} in equation (1) and on simplifying we then obtain

$$K_c = \frac{n_{NH_3}^2\, n^2 (RT)^2}{n_{N_2}\, n_{H_2}^3\, p^2} \qquad (9)$$

Equation (9) can be used for the calculation of K_c since T and p are given in the data, R is a known constant, and all the n values required could be calculated by making use of the balanced equation. The steps in this calculation are tabulated in Table 7.1 (if you find any difficulty in understanding any of these steps, study section 4.1 once again).

Table 7.1   Calculation for Example 7.3

Balanced equation	N_2	+	$3H_2$	$\rightarrow$	$2NH_3$
initial moles	1.00		3.00		0
moles reacted (because 30% reacts)	0.30		0.90		—
moles formed (by balanced equation)	—		—		0.60
moles present at equilibrium	$(1.00 - 0.30) = 0.70$		$(3.00 - 0.90) = 2.10$		0.60

That is, $n_{N_2} = 0.70$ mol, $n_{H_2} = 2.10$ mol and $n_{NH_3} = 0.60$ mol. The total moles n present in the equilibrium mixture is given by

$$n = n_{N_2} + n_{H_2} + n_{NH_3}$$

$$= 0.70 \text{ mol} + 2.10 \text{ mol} + 0.60 \text{ mol}$$

$$= 3.40 \text{ mol}$$

Step 4 Collect the data, check the units, and calculate
We can now calculate K_c by the application of equation (9).

$$n_{NH_3} = 0.60 \text{ mol}$$

$$n_{N_2} = 0.70 \text{ mol}$$

$$n_{H_2} = 2.10 \text{ mol}$$

$$n = 3.40 \text{ mol}$$

$$R = 8.314 \text{ J K}^{-1} \text{ mol}^{-1}$$

$$= 8.314 \text{ kg m}^2 \text{ s}^{-2} \text{ K}^{-1} \text{ mol}^{-1}$$

$$p = 3.55 \times 10^7 \text{ Pa} = 3.55 \times 10^7 \text{ kg m}^{-1} \text{ s}^{-2}$$

$$T = 773 \text{ K}$$

$$\therefore \quad K_c = \frac{(0.60 \text{ mol})^2 (3.40 \text{ mol})^2 (8.314 \text{ kg m}^2 \text{ s}^{-2} \text{ K}^{-1} \text{ mol}^{-1} \times 773 \text{ K})^2}{(0.70 \text{ mol}) (2.10 \text{ mol})^3 (3.55 \times 10^7 \text{ kg m}^{-1} \text{ s}^{-2})^2}$$

$$= 2.10 \times 10^{-8} \text{ mol}^{-2} \text{ m}^6$$

Solution to Part (b)
By following a procedure similar to that outlined in Part (a) it can be shown (check this as an exercise) that K_p is given by

$$K_p = \frac{n_{NH_3}^2 \, n^2}{n_{N_2} \, n_{H_2}^3 \, p^2}$$

$$= \frac{(0.60 \text{ mol})^2 (3.40 \text{ mol})^2}{(0.70 \text{ mol}) (2.10 \text{ mol})^3 (3.55 \times 10^7 \text{ Pa})^2}$$

$$= 5.10 \times 10^{-16} \text{ Pa}^{-2}$$

Step 5 Review and check the solution
The units are correct and the number of significant figures are reasonable.

Notes
1 Recognize that the problem of calculating an equilibrium constant is essentially a problem in calculating concentrations or partial pressures.
2 As for any other physical quantity, the units for K must always be given. The units for K follow directly from the defining equation for K. Conversely, the units given for an equilibrium constant would indicate the units that had been used for expressing concentrations or partial pressures.

Exercise 7.3

When 0.500 mole of phosphorus pentachloride is heated, 20.0% of it dissociates according to the equation $PCl_5(g) \rightleftharpoons PCl_3(g) + Cl_2(g)$. Calculate the moles of each gas present in the mixture.

Exercise 7.4

0.1772 g of disulphur dichloride (S_2Cl_2) occupy a volume 95.4 cm^3 at 798 K and at a pressure 9.96×10^4 Pa. Calculate the degree of dissociation and percentage dissociation. S_2Cl_2 dissociates according to the equation

$$S_2Cl_2(g) \rightleftharpoons S_2(g) + Cl_2(g)$$

Exercise 7.5

Write the expression for the equilibrium constant K_c for each of the following equilibria.

$$(1) \ C(s) + \tfrac{1}{2}O_2(g) \rightleftharpoons CO(g)$$
$$(2) \ CO(g) + \tfrac{1}{2}O_2(g) \rightleftharpoons CO_2(g)$$
$$(3) \ C(s) + O_2(g) \rightleftharpoons CO_2(g)$$

Derive the relationship between the equilibrium constants of (1), (2), and (3).

Exercise 7.6

Hydrogen and nitrogen in the molar ratio 3.00:1.00 respectively were maintained at 700 K and at pressure 1.00×10^5 Pa. The equilibrium mixture then contained 15.3% by volume of ammonia. Calculate K_p and K_c for the reaction

$$N_2(g) + 3H_2(g) \rightleftharpoons 2NH_3(g)$$

Exercise 7.7

Calculate K_c for the reaction $CO_2(g) + H_2(g) \rightleftharpoons CO(g) + H_2O(g)$ at 1173 K and 1.500×10^5 Pa if at equilibrium the partial pressures of CO_2, H_2, and CO respectively are 6.48×10^4 Pa, 1.48×10^4 Pa, and 3.52×10^4 Pa.

Exercise 7.8

3.00 g of phosphorus pentachloride (g) are heated in a 1.00 dm^3 vessel at 523 K until the following gaseous equilibrium

$$PCl_5(g) \rightleftharpoons PCl_3(g) + Cl_2(g)$$

is established. The equilibrium pressure, measured by a manometer attached to the vessel, is 1.15 atm. Calculate K_p for the reaction.

7.3 Importance of equilibrium constant

The equilibrium constant is very important because it summarizes a large amount of useful information about an equilibrium. Basically it shows quantitatively (see

Equation 7.1) the relationship that exists at equilibrium between the concentrations of the products and the concentrations of the reactants.

We have already emphasized (see section 1.4) that an understanding of the information provided by equations is necessary for understanding much of chemistry. Let us analyse **Equation 7.1** so as to extract from it the information it summarizes.

In **Equation 7.1**, K is a constant at a constant temperature. It follows therefore from the equation that there is always a definite relationship, at equilibrium, between the concentrations of the reactants and the concentrations of products. Using this equation we can deduce, provided K is known, the concentrations (and therefore amounts) of products that can be formed from known initial concentrations of reactants. That is, we can predict something very useful—the extent of a reaction or the yield of a reaction for any initial concentrations of reactants (this aspect is illustrated in Examples 7.5 and 7.7).

Equation 7.1 may also be used for deducing (both quantitatively and qualitatively) how the composition of a system at equilibrium depends on various factors such as

(a) the concentrations of reactants
(b) the concentrations of products
(c) the pressure (see Example 7.5)
(d) the volume.

Since **Equation 7.1** has only concentration terms it may appear that it cannot be used for deducing the effect of pressure or volume on the composition at equilibrium. This is not correct. A little thought, however, will make it clear that any factor that changes concentration will also affect the equilibrium composition of a system. For example, a change in pressure or volume of a gaseous system in equilibrium would change (immediately) the concentrations of the various substances present. The concentrations of the various substances would then adjust themselves (by chemical reaction in the appropriate direction) so as to satisfy **Equation 7.1**.

The effect of *changes* in various factors (concentration, pressure, volume) on the composition at equilibrium can also be deduced. For example, since K is a constant it follows (see **Equation 7.1**) that

(a) an increase in concentration of one (or more) of the reactants (A, B...) must increase the concentration of the products ...Y, Z (because the ratio $[Y]^y \times [Z]^z/[A]^a \times [B]^b$ must always be a constant at a constant temperature)
(b) the removal of a product (...Y, Z) from the equilibrium mixture must result in the further conversion of reactants into products (since K is a constant)
(c) the addition of a product of the reaction to an equilibrium mixture must result in the conversion of part of the products into reactants (see Example 7.7).

Example 7.4 Prediction of the yield from equilibrium constant data

The equilibrium constant for the homogeneous gas dissociation $N_2O_4(g) \rightleftharpoons 2NO_2(g)$ is 9.75×10^3 Pa at 300 K. Predict the moles of NO_2 molecules that would be formed at equilibrium if 0.1000 mol N_2O_4 is allowed to reach equilibrium at 300 K and 1.00×10^5 Pa.

Solution

Step 1 Clarify and define the problem

We are asked to calculate the moles of NO_2 (n_{NO_2}) present in the system from K_p data. To do so our objective should be to derive an equation that shows the relationship between K_p and n_{NO_2}.

Step 2 Select the key equation

We start with the defining equation for K_p which for the reaction considered is

$$K_p = \frac{p^2_{NO_2}}{p_{N_2O_4}} \qquad \text{(1) \textbf{Equation 7.1}}$$

Step 3 Derive the equation for the calculation

The problem now is: how could we insert the required quantity—n_{NO_2}—into equation (7)? We could replace the partial pressure terms in equation (1) by terms involving the number of moles by making use of **Equation 5.1**. By the application of this equation ($p_i = n_i RT/V$) we get

$$p_{NO_2} = \frac{n_{NO_2} RT}{V} \qquad \text{(2) \textbf{Equation 5.1}}$$

$$p_{N_2O_4} = \frac{n_{N_2O_4} RT}{V} \qquad \text{(3) \textbf{Equation 5.1}}$$

On substituting equations (2) and (3) in equatioin (1) and on simplifying we obtain

$$K_p = \frac{n^2_{NO_2} RT}{n_{N_2O_4} V} \qquad \text{(4)}$$

The volume term in equation (4) is not given in the data; p, however, is given. We therefore replace V by nRT/p (see **Equation 5.1**) when we obtain

$$K_p = \frac{n^2_{NO_2} p}{n_{N_2O_4} n} \qquad \text{(5)}$$

Consider now the terms in equation (5); K_p and p are given in the data. n_{NO_2}, $n_{N_2O_4}$, and n (total moles) are unknown; these quantities are, however, related to each other by the balanced equation for the reaction and it is therefore possible to calculate the required quantity n_{NO_2} as is shown below, by making use of equation (5).

Let the required quantity n_{NO_2} be denoted by x. From the chemical equation given in the problem it can be shown that (see note below)

$$n_{N_2O_4} = (0.1000 - 0.500x) \text{ mol}$$

$$n = (0.1000 + 0.500x) \text{ mol}$$

Step 4 Collect the data, check the units, and calculate

The data given are: K_p 9.75 × 10^3 Pa, and p 1.00 × 10^5 Pa. Expressions for n_{NO_2}, $n_{N_2O_4}$, and n were obtained in Step 3. Equation (5) can therefore be written as

$$9.75 \times 10^3 \text{ Pa} = \frac{(x^2 \text{ mol}^2) \times (1.00 \times 10^5 \text{ Pa})}{(0.1000 - 0.500x) \text{ mol} (0.1000 + 0.500x) \text{ mol}} \tag{6}$$

On simplifying the above equation we obtain

$$1.024x^2 = 9.75 \times 10^{-4} \text{ mol}^2$$

$$\therefore \qquad x = 3.08 \times 10^{-2} \text{ mol}$$

The moles of NO_2 molecules that would be formed at equilibrium is therefore 3.08×10^{-2} mol.

Step 5   Review and check the solution

The units in the answer are correct and it appears reasonable.

Note

The moles of N_2O_4 present after dissociation $n_{N_2O_4}$ is given by

$$n_{N_2O_4} = \text{initial moles} - \text{moles dissociated}$$

$$= 0.1000 \text{ mol} - 0.500x \text{ mol}$$

(moles dissociated $= 0.500x$ mol; because the balanced equation shows that to form x mol NO_2, $0.500x$ mol of N_2O_4 must dissociate).

The total moles n present in the equilibrium mixture is given by

$$n = n_{N_2O_4} + n_{NO_2}$$

$$= (0.1000 - 0.500x) \text{ mol} + x \text{ mol}$$

$$= (0.1000 + 0.500x) \text{ mol}$$

Example 7.5   Effect of pressure on the yield of a reaction

Suppose that the pressure of the equilibrium system ($N_2O_4 \rightleftharpoons 2NO_2$) in Example 7.4 is increased (from 1.00×10^5 Pa) to 2.00×10^5 Pa. Deduce the effect of this pressure increase on the yield of NO_2.

Solution

Step 1   Clarify and define the problem

To answer this question we should know the yields in the initial state (i.e. when $p = 1.00 \times 10^5$ Pa) and in the final state (i.e. when $p = 2.00 \times 10^5$ Pa). In Example 7.4 it was shown that the yield at 1.00×10^5 Pa is 3.08×10^{-2} mol. So we need to calculate here only the yield when $p = 2.00 \times 10^5$ Pa.

The method of calculation is the same as that in Example 7.4; only a brief outline of the solution will therefore be given.

Steps 2 and 3   Key equation and equation for the calculation

These are the same as those given in 7.4.

Step 4   Collect the data, check the units, and calculate

The only difference from the calculation in Example 7.4 is that here $p = 2.00 \times 10^5$ Pa (instead of 1.00×10^5 Pa). Substitution of this value for p into equation (6) of Example 7.4 (instead of 1.00×10^5 Pa) gives

$$9.75 \times 10^3 \text{ Pa} = \frac{x^2 \times 2.00 \times 10^5 \text{ Pa}}{(0.1000 - 0.500x) \text{ mol} (0.1000 + 0.500x) \text{ mol}}$$

On simplification we obtain

$$2.024x^2 = 9.75 \times 10^{-4} \text{ mol}^2$$

$$\therefore \qquad x = 2.20 \times 10^{-2} \text{ mol}$$

When p is increased from 1.00×10^5 Pa to 2.00×10^5 Pa the yield is therefore decreased. The decrease in yield is given by

$$\text{decrease in yield} = \text{yield at } 1.00 \times 10^5 \text{ Pa} - \text{yield at } 2.00 \times 10^5 \text{ Pa}$$

$$= 3.08 \times 10^{-2} \text{ mol} - 2.20 \times 10^{-2} \text{ mol}$$

$$= 8.8 \times 10^{-3} \text{ mol}$$

Step 5 Review and check the solution

The above example indicates how the effect of pressure on the yield could be deduced quantitatively by making use of the equation for the equilibrium constant. The answer shows that there is a decrease in yield; this can be checked by applying Le Chatelier's principle.

Le Chatelier's principle states that if a system in equilibrium is disturbed (for example, by changing the pressure, temperature, or concentrations of the participating substances) the composition of the system will change in such a way as to oppose the effect of the disturbance.

In this example ($N_2O_4 \rightleftharpoons 2NO_2$) the pressure of the system is increased. According to Le Chatelier's principle the composition of the system should change in a manner that would reduce the pressure. Pressure could be reduced by a reduction in the moles n in the equilibrium system (since $p \propto n$, see **Equation 5.1**). The reaction which is accompanied by a reduction in the moles (which is the formation of N_2O_4 from NO_2) would therefore take place. Le Chatelier's principle therefore predicts that an increase in pressure would favour the formation of N_2O_4 from NO_2, i.e. it would decrease the yield of NO_2.

The effect of a pressure change on the composition at equilibrium could similarly be predicted for any other equilibrium system; qualitatively by Le Chatelier's principle or quantitatively by the equation for the equilibrium constant.

Example 7.6 Calculation of degree of dissociation from K_c data

Calculate the degree of dissociation when 0.500 mol phosphorus pentachloride is maintained at 573 K and 2.00×10^5 Pa. The equilibrium constant for this homogeneous gas dissociation, $PCl_5(g) \rightleftharpoons PCl_3(g) + Cl_2(g)$, is 2.30×10^{-3} mol dm^{-3} at 573 K.

Solution

Step 1 Clarify and define the problem

We have to calculate the degree of dissociation (α) from K_c data. Our objective should, therefore, be to derive an equation that shows the relationship between K_c and α.

Step 2 Select the key equation

We start with the defining equation for K_c for the example considered which is

$$K_c = \frac{c_{PCl_3} c_{Cl_2}}{c_{PCl_5}}$$

(1) **Equation 7.1**

Step 3 Derive the equation for the calculation

We want to introduce α into equation (1). This could be done because the concentration terms in the equation could be given in terms of α by making use of the balanced equation for the dissociation. The steps in this calculation are tabulated in Table 7.2.

Table 7.2   Calculation for Example 7.6

Balanced equation	PCl_5	$\rightleftharpoons$	PCl_3	+	Cl_2
initial moles	0.500		0		0
moles dissociated (by **equation 2.2**)	0.500α				
moles formed (by balanced equation)			0.500α		0.500α
moles at equilibrium	$(0.500 - 0.500\alpha)$		0.500α		0.500α
	$= 0.500(1 - \alpha)$				
concentration c	$\dfrac{0.500(1 - \alpha)}{V}$		$\dfrac{0.500\alpha}{V}$		$\dfrac{0.500\alpha}{V}$

On substituting the expressions for the concentrations obtained above in equation (1), and on simplifying we obtain

$$K_c = \frac{0.500\alpha^2}{(1 - \alpha) V}$$

(2)

Equation (2) relates K_c and α. To calculate α using the equation V is, however, required. This is not given in the data; p is given. We therefore replace V by nRT/p (see **Equation 5.1**) when we get

$$K_c = \frac{0.500\alpha^2 p}{(1 - \alpha) nRT}$$

(3)

In equation (3)

$$n = \text{total number of moles at equilibrium}$$

$$= n_{PCl_5} + n_{PCl_3} + n_{Cl_2}$$

$$= 0.500(1 - \alpha) + 0.500\alpha + 0.500\alpha$$

$$= 0.500(1 + \alpha)$$

(4)

On substituting equation (4) in (3) and on simplifying we get

$$K_c = \frac{\alpha^2 p}{RT(1 - \alpha)(1 + \alpha)} = \frac{\alpha^2 p}{RT(1 - \alpha^2)}$$

$$\therefore \quad \frac{\alpha^2}{1 - \alpha^2} = \frac{K_c RT}{p}$$

(5)

Step 4 Collect the data, check the units, and calculate

α could be calculated using equation (5) since it is the only unknown.

$$K_c = 2.30 \times 10^{-3} \, \text{mol dm}^{-3} = 2.30 \, \text{mol m}^{-3}$$

$$R = 8.314 \, \text{kg m}^2 \, \text{s}^{-2} \, \text{K}^{-1} \, \text{mol}^{-1}$$

$$T = 573 \, \text{K}$$

$$p = 2.00 \times 10^5 \, \text{Pa} = 2.00 \times 10^5 \, \text{kg m}^{-1} \, \text{s}^{-2}$$

$$\frac{\alpha^2}{1 - \alpha^2} = \frac{(2.30 \, \text{mol m}^{-3}) \, (8.314 \, \text{kg m}^2 \, \text{s}^{-2} \, \text{K}^{-1} \, \text{mol}^{-1}) \, (573 \, \text{K})}{2.00 \times 10^5 \, \text{kg m}^{-1} \, \text{s}^{-2}}$$

$$= 5.48 \times 10^{-2}$$

$$\therefore \qquad \alpha = 0.228$$

The required degree of dissociation is therefore 0.228.

Step 5 Review and check the solution

The answer is reasonable because by definition the degree of dissociation must lie between 0 and 1.

Example 7.7 Effect on yield of the addition of a product of a reaction

Consider Example 7.6. What would be the degree of dissociation if initially there had been 0.500 mol PCl_5 and 0.500 mol Cl_2, instead of pure PCl_5?

Solution

Step 1 Clarify and define the problem

The method of calculation is similar to that given in Example 7.6.

Step 2 Select the key equation

The key equation is the same as that in Example 7.6.

Step 3 Derive the equation for calculation

The method is the same as that in Example 7.6. The only difference here is that the expression for the moles of Cl_2 at equilibrium (n_{Cl_2}) would be seen to have a different value which is

$$n_{Cl_2} = (0.500 + 0.500\alpha)$$

If this expression for n_{Cl_2} is used (instead of $n_{Cl_2} = 0.500\alpha$ as was used for Example 7.6), the final equation for the calculation (instead of equation 5 in Example 7.6) would be given by

$$\frac{1 + \alpha}{(1 - \alpha) (2 + \alpha)} = \frac{K_c RT}{p} \qquad (1)$$

Step 4 Collect the data, check the units, and calculate

It was shown in Example 7.6 that for the example being considered $K_c RT/p$ is equal to 5.48×10^{-2}. Equation (1) therefore becomes

$$\frac{1 + \alpha}{(1 - \alpha)(2 + \alpha)} = 5.48 \times 10^{-2}$$

On simplification the above equation becomes

$$1.055\alpha^2 + 1.055\alpha - 0.1096 = 0$$

On solving this *quadratic equation* (for method of solution see note below) we get

$$\alpha = 0.095$$

The degree of dissociation is therefore 0.095.

Step 5 *Review and check the solution*

Comparison of the answers in Examples 7.6 and 7.7 shows that the answers are reasonable because the degree of dissociation is seen to decrease in the presence of a product of the reaction (chlorine in this example). This decrease is also to be expected from Le Chatelier's principle.

Note on solution of quadratic equations

In this example we had to solve the equation

$$1.055\alpha^2 + 1.055\alpha - 0.1096 = 0$$

This is a quadratic equation (a quadratic equation is one in which the highest power of the variable is 2; in the above equation the highest power of the variable α is seen to be 2). Two methods for solving a quadratic equation are outlined now.

(a) One method of solving a quadratic equation is to first rearrange it into the form

$$ax^2 + bx + c = 0 \qquad (1)$$

where x is the variable and a, b, and c are constants. The solution is then given by the equation

$$x = \frac{-b \pm (b^2 - 4ac)^{1/2}}{2a} \qquad (2)$$

The solution of the equation

$$1.055\alpha^2 + 1.055\alpha - 0.1096 = 0 \qquad (3)$$

by the application of equation (2) will now be illustrated. Comparison of equations (1) and (3) shows that $a = 1.055$, $b = 1.055$, and $c = -0.1096$. On applying equation (2) to solve equation (3) we get

$$\alpha = \frac{-1.055 \pm \{1.055^2 - 4 \times 1.055 \times (-0.1096)\}^{1/2}}{2 \times 1.055}$$

$$= \frac{-1.055 \pm 1.256}{2.110}$$

$$= +0.095 \quad \text{or} \quad -1.095$$

There are thus 2 answers—one positive and one negative. The definition shows that α cannot have a negative value. The negative answer should therefore be ignored.

$$\therefore \qquad \alpha = 0.095$$

(b) Another method of solving a quadratic equation, which is particularly useful when x is very small, is the *method of successive approximations*. Consider equation (3)

$$1.055\alpha^2 + 1.055\alpha - 0.1096 = 0 \tag{4}$$

If α is small it can be assumed, as a first approximation, that α^2 would be negligibly small compared with α. Equation (3) could then be written as

$$1.055\alpha - 0.1096 = 0$$

$$\therefore \qquad \alpha = 0.104$$

The value of α so calculated cannot, of course, be accurate because it is based on the assumption that the α^2 term in equation (4) is zero. A more accurate value for α could, however, be obtained by inserting this value of α (0.104) for calculating the α^2 term in equation (4). Equation (4) then becomes

$$1.055 \times (0.104)^2 + 1.055\alpha - 0.1096 = 0$$

from which $\alpha = 0.093$.

A still more accurate value for α could be obtained by inserting the last value of α (0.093) for calculating the α^2 term in equation (4). Then we get

$$1.055 \times (0.093)^2 + 1.055\alpha - 0.1096 = 0$$

from which $\alpha = 0.095$. Insertion of this value of α into the α^2 term in equation (4) and calculation of α shows that it has the same value (0·095). This is therefore the correct answer.

Three successive approximations were required to obtain the correct answer in this calculation. In this case it may therefore be easier to use equation (2) to solve the quadratic equation.

When α is very small (less than 0.05) one approximation, however, is generally sufficient to give the correct answer. The method of successive approximation would then give the solution more easily.

Exercise 7.9

Calculate the degree of dissociation of SO_3 at 1.52×10^5 Pa if K_p for the reaction $SO_3(g) \rightleftharpoons SO_2(g) + \frac{1}{2}O_2(g)$ is 3.14×10^{-3} $Pa^{-1/2}$.

Exercise 7.10

NOCl dissociates on heating into NO and Cl_2. If 1.00 mol NOCl(g) is heated at 473 K, at what total pressure would the partial pressure of Cl_2 be 1.21×10^4 Pa? $K_p = 2.22 \times 10^2$ Pa at 473 K.

Exercise 7.11

When 0.100 mol $PCl_5(g)$ is heated in a 100 cm^3 vessel at 500 K, 13.9% dissociates into $PCl_3(g)$ and $Cl_2(g)$. Calculate the decrease in the degree of dissociation if 0.050 mol of Cl_2 is injected into the equilibrium mixture. Assume that the volume is kept constant. The equilibrium constant for the dissociation is 2.24×10^{-2} mol dm^{-3} at 500 K.

Exercise 7.12

Consider the following reactions: (a) $PCl_5(g) \rightleftharpoons PCl_3(g) + Cl_2(g)$ (b) $H_2(g) + I_2(g) \rightleftharpoons 2HI(g)$. By making use of the law of chemical equilibrium deduce, in each case, the effect of total pressure on the yield of product.

Exercise 7.13

Consider the reaction $2SO_2(g) + O_2(g) \rightleftharpoons 2SO_3(g)$. Derive an equation that could be used to predict the effect of (a) total pressure (b) volume, on the yield of SO_3. Will an increase in pressure increase or decrease the yield of SO_3?

Exercise 7.14

0.400 mol of $N_2O_4(g)$ and 0.800 mol of $NO_2(g)$ are present in equilibrium at 50 °C in a 2.00 dm^3 vessel. Calculate the moles of N_2O_4 molecules that will be present in the system if

    (a) the volume of the system is doubled
    (b) the pressure of the system is doubled
    (c) 0.400 mol of NO_2 is introduced at constant volume
    (d) 0.400 mol of NO_2 is introduced at constant pressure
    (e) 0.400 mol of an inert gas is introduced at constant volume
    (f) 0.400 mol of an inert gas is introduced at constant pressure.

Example 7.8 Calculation of K_c from K_p

K_p for the reaction $2SO_2(g) + O_2(g) \rightleftharpoons 2SO_3(g)$ is 0.057 Pa^{-1} at 800 K. Calculate K_c.

Solution

Step 1 Clarify and define the problem
To calculate K_c from K_p data we must first derive the equation relating them.

Step 2 Select the key equation
We start with the defining equation for the required quantity K_c which for the equilibrium considered is

$$K_c = \frac{c_{SO_3}^2}{c_{SO_2}^2 c_{O_2}} \qquad (1)$$

Step 3 Derive the equation for calculation
Since we have to relate K_c and K_p, we now write the defining equation for K_p which is

$$K_p = \frac{p_{SO_3}^2}{p_{SO_2}^2 p_{O_2}} \qquad (2)$$

Equations (1) and (2) show that to relate K_c and K_p we must know the relationship between p and c for a gas. This is easily obtained by the combination of the equations

$$c = n/V \qquad\qquad \textbf{Equation 4.2}$$

and

$$p = nRT/V \qquad\qquad \textbf{Equation 5.1}$$

when we obtain

$$c = \frac{p}{RT} \qquad\qquad (3)$$

Application of equation (3) to each of the gases present in the equilibrium mixture gives

$$c_{SO_3} = \frac{p_{SO_3}}{RT}$$

$$c_{SO_2} = \frac{p_{SO_2}}{RT}$$

$$c_{O_2} = \frac{p_{O_2}}{RT}$$

On replacing the concentration terms in equation (1) by the above equations we then obtain

$$K_c = \frac{(p_{SO_3}/RT)^2}{(p_{SO_2}/RT)^2\,(p_{O_2}/RT)}$$

$$= \frac{p_{SO_3}^2}{p_{SO_2}^2\,p_{O_2}} \times (RT)$$

$$= K_p \times (RT) \qquad\qquad (4)$$

Step 4 Collect the data, check the units, and calculate

We can now calculate K_c by substituting the given data in equation (4).

$$K_p = 0.057\ \text{Pa}^{-1} = 0.057\ \text{kg}^{-1}\ \text{m}\ \text{s}^2$$

$$T = 800\ \text{K}$$

$$K_c = (0.057\ \text{kg}^{-1}\ \text{m}\ \text{s}^2) \times (8.314\ \text{kg}\ \text{m}^2\ \text{s}^{-2}\ \text{K}^{-1}\ \text{mol}^{-1} \times 800\ \text{K})$$

$$= 3.79 \times 10^2\ \text{mol}^{-1}\ \text{m}^3$$

Step 5 Review and check the solution

The units for K_c are correct (check by applying **Equation 7.1**) and the number of significant figures reasonable.

Example 7.9 Equilibrium constants for the same equilibrium written in different ways

The equilibrium constant K_p for the reaction $SO_2(g) + \frac{1}{2}O_2(g) \rightleftharpoons SO_3(g)$ is 3.0×10^{-3} $Pa^{-1/2}$ at 1000 K. Calculate K_p for the reactions

(a) $SO_3(g) \rightleftharpoons SO_2(g) + \frac{1}{2}O_2(g)$
(b) $2SO_2(g) + O_2(g) \rightleftharpoons 2SO_3(g)$.

Solution to Part (a)

Step 1   Clarify and define the problem

The equilibrium constant K_p for the reaction

$$SO_3 \rightleftharpoons SO_2 + \tfrac{1}{2}O_2 \tag{a}$$

is required. Data give the equilibrium constant K for the reaction

$$SO_2 + \tfrac{1}{2}O_2 \rightleftharpoons SO_3 \tag{b}$$

The unit ($Pa^{1/2}$) given for K indicates that it is in terms of partial pressures; let us denote the equilibrium constant for reaction (b) by K_p'.

Our problem is to calculate K_n (equilibrium constant for reaction a) from K_p' (equilibrium constant for reaction b, this is given in the data). To do so we must derive an equation which relates K_p and K_p'.

Step 2   Select the key equation

K_p for reaction (a) is required; we therefore start with the defining equation for this quantity which is

$$K_p = \frac{p_{SO_2} p_{O_2}^{1/2}}{p_{SO_3}} \tag{1} \textbf{ Equation 7.1}$$

Step 3   Derive the equation for the calculation

The right side of equation (1) can be related to the datum given (K_p'). Since K_p' is defined by the equation

$$K_p' = \frac{p_{SO_3}}{p_{SO_2} p_{O_2}^{1/2}} \tag{2}$$

it follows that

$$\frac{p_{SO_3} p_{O_2}^{1/2}}{p_{SO_3}} = \frac{1}{K_p'} \tag{3}$$

Combination of equations (1) and (3) then gives

$$K_p = \frac{1}{K_p'} \tag{4}$$

Equation (4) can be used to calculate the required quantity (K_p).

Step 4 Collect the data, check the units, and calculate

$$K'_p = 3.0 \times 10^{-3} \text{ Pa}^{-1/2}$$

$$\therefore \quad K_p = \frac{1}{3.0 \times 10^{-3} \text{ Pa}^{-1/2}} = \frac{1 \times 10^3 \text{ Pa}^{1/2}}{3.0}$$

$$= 3.3 \times 10^2 \text{ Pa}^{1/2}$$

The equilibrium constant for the reaction $SO_3(g) \rightleftharpoons SO_2(g) + \frac{1}{2}O_2(g)$ is therefore $3.3 \times 10^2 \text{ Pa}^{1/2}$.

Solution to Part (b)

Step 1 Clarify and define the problem

We have to calculate K_p for the reaction

$$2SO_2 + O_2 \rightleftharpoons 2SO_3 \tag{5}$$

Step 2 Select the key equation

We start with the defining equation for the required quantity (K_p) which is

$$K_p = \frac{p^2_{SO_3}}{p^2_{SO_2} p_{O_2}} \tag{6} \textbf{ Equation 7.1}$$

Step 3 Derive the equation for the calculation

The right side of equation (6) can be related to K'_p by equation (2). If we square equation (2) we obtain

$$\frac{p^2_{SO_3}}{p^2_{SO_2} p_{O_2}} = K'^2_p \tag{7}$$

Combination of equations (6) and (7) then shows that

$$K_p = K'^2_p \tag{8}$$

The required K_p can therefore be calculated.

Step 4 Collect the data, check the units, and calculate

$$K'_p = 3.0 \times 10^{-3} \text{ Pa}^{-1/2}$$

$$\therefore \quad K_p = (3.0 \times 10^{-3} \text{ Pa}^{-1/2})^2$$

$$= 9.0 \times 10^{-6} \text{ Pa}^{-1}$$

The equilibrium constant for the reaction $2SO_2(g) + O_2(g) \rightleftharpoons 2SO_3(g)$ is therefore $9.0 \times 10^{-6} \text{ Pa}^{-1}$.

Step 5 Review and check the solutions

Checking with the corresponding balanced equation shows that the units of the answers are correct. The number appearing as an exponent in the units should always be the same as Δn, the change in the number of moles during the reaction considered. In Part (b), for example, equation (5) shows that $\Delta n = -1$; the exponent in the units of the answer should therefore be -1. This is found to be so (Pa^{-1}).

Exercise 7.15

For the equilibrium between carbon monoxide (CO), oxygen (O_2), and carbon dioxide (CO_2) an equilibrium constant has the units $mol^{1/2} \ dm^{-3/2}$. Write the equation corresponding to this equilibrium constant. In what unit had the concentrations been given?

Exercise 7.16

At 1000 K, the equilibrium constant for the reaction $N_2(g) + 3H_2(g) \rightleftharpoons 2NH_3(g)$ is 2.37×10^{-3} $mol^{-2} \ dm^6$. Deduce the equilibrium constants for the following reactions:

(a) $2NH_3(g) \rightleftharpoons N_2(g) + 3H_2(g)$
(b) $NH_3(g) \rightleftharpoons \frac{1}{2}N_2(g) + \frac{3}{2}H_2(g)$.

Exercise 7.17

Derive the equation showing the relationship between K_p and K_c for the reaction $N_2(g) + 3H_2(g) \rightleftharpoons 2NH_3(g)$.

Review of Chapter 7

Only one equation was introduced in this chapter; the equation for the law of chemical equilibrium which for a general reaction, $aA + bB + ... \rightleftharpoons ...yY + zZ$ is written as

$$K = \frac{... \times [Y]^y \times [Z]^z}{[A]^a \times [B]^b \times ...}$$ 7.1

K is the equilibrium constant. The units for K depend on the stoichiometry of the equation. This equation is of very wide applicability; it applies to *all* types of equilibria.

Equation 7.1 is very important because it provides a basis for the understanding of the behaviour of all types of systems in equilibrium. By the application of this equation we can predict both qualitatively and quantitatively the influence of various factors (concentration of reactants, concentrations of products, pressure, volume) on the composition of any equilibrium.

Application of this equation is not difficult. Difficulties sometimes encountered by students are associated mainly with the concentration terms in this equation—how to calculate the concentrations (molar concentrations, partial pressure) of the various substances *at equilibrium*, how to convert concentrations from one unit to another, how to calculate concentrations after partial chemical reaction, etc. These aspects were explained in Chapter 4 (particularly in section 4.4).

Calculations involving equilibria are often simplified by tabulating the data and the steps in the calculation. Tabulation presents the problem and the solution clearly.

8 | Ionic equilibria

It was emphasized in Chapter 7 that the law of chemical equilibrium is a very general and important law. It applies to all types of equilibria. This chapter illustrates the application of this law to various ionic equilibria in *aqueous* solutions.

8.1 Dissociation of water and pH

Water dissociates according to the equation

$$H_2O(l) \rightleftharpoons H^+(aq) + OH^-(aq)$$

The concentration of water molecules can be considered to remain constant since the number of these molecules which dissociates is negligibly small compared with the number that remains undissociated. The equilibrium constant for the dissociation, which is called the *ionic product* and is denoted by K_w, is therefore given by the application of **Equation 7.1** as

$$K_w = [H^+][OH^-]$$

As usual, square brackets [ ] denote concentrations.

It is more convenient to express the values of $[H^+]$, $[OH^-]$, and K_w as pH, pOH, and pK_w (see section 2.3). These are defined by

$$pH = -\log_{10}[H^+]$$

$$pOH = -\log_{10}[OH^-]$$

$$pK_w = -\log_{10} K_w$$

In these equations, $[H^+]$, $[OH^-]$, and K_w must be given in the units mol dm^{-3}, mol dm^{-3} and mol^2 dm^{-6} respectively*.

The three equations given above can be represented by the general equation

$$pX = -\log_{10} X \qquad \qquad \textbf{Equation 8.1}$$

where X is any physical property.

Example 8.1 Calculation of pH

The hydrogen ion concentration in an aqueous solution at 298 K is 2.80×10^{-5} mol dm^{-3}. At 298 K the value of K_w is 1.00×10^{-14} mol^2 dm^{-6}.

(a) Calculate the pH of the solution.
(b) Is this solution acidic or alkaline?

---

* Logarithms can be obtained only for a *number*; it is not possible to obtain the logarithm of a *unit*. If we are to have a unique value for the logarithm of a particular physical quantity it is, therefore, necessary to specify the unit in which that physical quantity is given.

Solution to Part (a)

To calculate pH we start with the defining equation for pH which by the application of **Equation 8.1** is

$$pH = -\log_{10}[H^+] \qquad \text{Equation 8.1}$$

This equation shows that to calculate pH we must know the hydrogen ion concentration $[H^+]$ of the solution. Since the hydrogen ion concentration is given in the data, the pH is easily calculated by the direct application of this equation.

$$pH = -\log_{10}[H^+]$$
$$= -\log_{10}(2.80 \times 10^{-5})$$
$$= -(\log_{10} 2.80 + \log_{10} 10^{-5})$$
$$= -(0.4472 - 5)$$
$$= 4.55$$

Solution to Part (b)

Step 1 Clarify and define the problem

We have to find out whether the given solution is acidic or alkaline. To do so we make use of the concept that an aqueous solution is acidic if $[H^+] > [OH^-]$, and that it is alkaline if $[OH^-] > [H^+]$. To decide whether a particular solution is acidic or alkaline we should therefore know the concentrations of H^+ and OH^- in the solution.

$[H^+]$ is given in the data. We have therefore to calculate only $[OH^-]$.

Step 2 Select the key equation

We have to calculate $[OH^-]$ in the solution. This could be done by the application of the **Equation 7.1** to the dissociation $H_2O(l) \rightleftharpoons H^+(aq) + OH^-(aq)$. We then obtain:

$$K_w = [H^+][OH^-] \qquad \text{Equation 7.1}$$

because the concentration of $H_2O(l)$ is constant.

$$\therefore \qquad [OH^-] = \frac{K_w}{[H^+]} \qquad (1)$$

Step 3 Derive the equation for the calculation

Equation (1) could be used for the calculation of $[OH^-]$ since K_w and $[H^+]$ are given in the data.

Step 4 Collect the data, check the units, and calculate

We are given that

$$K_w = 1.00 \times 10^{-14} \text{ mol}^2 \text{ dm}^{-6}$$
$$[H^+] = 2.80 \times 10^{-5} \text{ mol dm}^{-3}$$

$$\therefore \qquad [OH^-] = \frac{1.00 \times 10^{-14} \text{ mol}^2 \text{ dm}^{-6}}{2.80 \times 10^{-5} \text{ mol dm}^{-3}}$$

$$= 3.57 \times 10^{-10} \text{ mol dm}^{-3}$$

It is seen that in the solution given the hydrogen ion concentration (2.80×10^{-5} mol dm^{-3}) is greater than the hydroxyl ion concentration (3.57×10^{-10} mol dm^{-3}). The solution is therefore acidic.

Step 5 Review and check the solution

Note

Students often make the mistake of assuming that an aqueous solution is acidic if its hydrogen ion concentration is greater than 1.00×10^{-7} mol dm^{-3} (i.e. if pH < 7.00). This is only true at 25 °C (see Exercise 8.2 and its answer).

Example 8.2 Calculation of pH from ionic product

Calculate the pH of water at 298.2 K if the ionic product of water is 1.00×10^{-14} mol^2 dm^{-6} at this temperature.

Solution

Step 1 Clarify and define the problem

We have to calculate the pH of water from the ionic product K_w. To do so we must derive the equation showing the relation between pH and K_w.

Step 2 Select the key equation

We start with the defining equation for the required quantity (pH) which is

$$pH = -\log_{10} [H^+] \qquad \text{(1) Equation 8.1}$$

Step 3 Derive the equation for calculation

To calculate pH by equation (1) we require the hydrogen ion concentration [H$^+$] in the solution. This is not given in the data. Can we relate [H$^+$] to the data given (K_w)? Yes, by the defining equation for the ionic product of water from which

$$[H^+] = \frac{K_w}{[OH^-]} \qquad \text{(2) Equation 7.1}$$

To calculate [H$^+$], [OH$^-$] must therefore be known. Since we are dealing with pure water it is clear that

$$[OH^-] = [H^+] \qquad (3)$$

since the dissociation of water ($H_2O(l) \rightleftharpoons H^+(aq) + OH^-(aq)$) produces equal amounts of H$^+$ and OH$^-$ ions. On combining equations (2) and (3) we obtain

$$[H^+] = \frac{K_w}{[H^+]}$$

$$\therefore \qquad [H^+] = K_w^{1/2} \qquad (4)$$

On substituting for [H$^+$] in equation (1) by equation (4) we then obtain

$$pH = -\log_{10} K_w^{1/2}$$
$$= -\tfrac{1}{2}\log_{10} K_w \qquad (5)$$

Step 4   Collect the data, check the units, and calculate

$$K_w = 1.00 \times 10^{-14} \text{ mol}^2 \text{ dm}^{-6} \qquad \text{(from data)}$$

The numerical value for K_w can be inserted into equation (5) for the calculation since the units for K_w are in $\text{mol}^2 \text{ dm}^{-6}$ (see section 8.1).

$$\therefore \qquad \text{pH} = -\tfrac{1}{2} \log_{10} (1.00 \times 10^{-14})$$

$$= 7.00$$

Step 5   Review and check the solution

Exercise 8.1

The pH of a blood sample is 7.4 at 298 K. Calculate the hydrogen and hydroxyl ion concentrations. $K_w = 1.00 \times 10^{-14} \text{ mol}^2 \text{ dm}^{-6}$ at 298 K.

Exercise 8.2

An aqueous solution has a pH 6.8 at 333 K. Is this solution acidic or alkaline? $K_w = 9.55 \times 10^{-14} \text{ mol}^2 \text{ dm}^{-6}$ at 333 K.

8.2   Aqueous solutions of acids and bases

Acids, bases, and many salts affect the equilibrium due to the dissociation of water ($H_2O(l) \rightleftharpoons H^+(aq) + OH^-(aq)$) either by adding or by removing H^+ or OH^- ions.

Strong acids (e.g. HNO_3, HCl) and strong bases (e.g. NaOH, KOH) are fully dissociated in aqueous solution. Weak acids (e.g. CH_3COOH), in contrast, are only partially dissociated in aqueous solution; an equilibrium then exists between undissociated molecules and ions.

For a monobasic weak acid HA this equilibrium can be represented as

$$HA(aq) \rightleftharpoons H^+(aq) + A^-(aq)$$

The equilibrium constant K_a for this dissociation is commonly known as the *dissociation constant* and is given by

$$K_a = \frac{[H^+][A^-]}{[HA]}$$

A dibasic weak acid (e.g. $H_2C_2O_4$) or a tribasic weak acid (e.g. H_3PO_4) dissociates in stages. For example, a dibasic weak acid H_2A dissociates in two steps as follows

$$H_2A(aq) \rightleftharpoons H^+(aq) + HA^-(aq)$$

$$HA^-(aq) \rightleftharpoons H^+(aq) + A^{2-}(aq)$$

The *first dissociation constant* K_1 is the equilibrium constant for the first dissociation and therefore:

$$K_1 = \frac{[H^+][HA^-]}{[H_2A]}$$

The *second dissociation constant* K_2 is the equilibrium constant for the second dissociation and therefore:

$$K_2 = \frac{[H^+][A^{2-}]}{[HA^-]}$$

Values of K_a are often expressed as pK_a which is defined by **Equation 8.1** as

$$pK_a = -\log_{10} K_a$$

Example 8.3 Calculation of the hydrogen ion concentration in a weak acid solution

Calculate the hydrogen ion concentration of a 0.100 M aqueous solution of ethanoic acid at 25 °C. The dissociation constant of ethanoic acid at this temperature is 1.74 × 10^{-5} mol dm^{-3}.

Solution

Step 1 Clarify and define the problem

Let the required quantity—the hydrogen ion concentration in the solution—be denoted by $[H^+]$. We have to calculate $[H^+]$ from dissociation constant (K_a). To do so, we have to obtain an equation which shows the relation between K_a and $[H^+]$.

Step 2 Select the key equation

The required relation between K and $[H^+]$ is given by the defining equation for K_a. For the dissociation considered ($CH_3COOH(aq) \rightleftharpoons CH_3COO^-(aq) + H^+(aq)$), application of **Equation 7.1** shows that K_a is given by

$$K_a = \frac{[CH_3COO^-][H^+]}{[CH_3COOH]} \tag{1}$$

Step 3 Derive the equation for calculation

To calculate $[H^+]$ using equation (1), we must know $[CH_3COO^-]$ and $[CH_3COOH]$ at equilibrium.

These two concentrations are, however, related to $[H^+]$. The balanced equation for the dissociation ($CH_3COOH(aq) \rightleftharpoons CH_3COO^-(aq) + H^+(aq)$) shows that

$$[CH_3COO^-] = [H^+] \tag{2}$$

The balanced equation for the dissociation also shows that

$$[CH_3COOH] = c - [H^+] \tag{3}$$

where c is the initial concentration of acid.

On substituting equations (2) and (3) in equation (1) we then obtain

$$K_a = \frac{[H^+][H^+]}{c - [H^+]}$$

$$= \frac{[H^+]^2}{c - [H^+]} \tag{4}$$

Look at equation (4) and the data given. K_a and c are given in the data and the only unknown in the equation is the hydrogen ion concentration, $[H^+]$. It is therefore possible to substitute the values for K_a and c and calculate $[H^+]$. To do this calculation we must know the method of solving a quadratic equation (see note after Example 7.7).

We can, however, solve equation (4) and obtain $[H^+]$ much more easily by making an approximation. Since K_a is small it means that $[H^+]$ too would be small. We may therefore assume that $[H^+]$ would be negligibly small compared with c and so write

$$c - [H^+] = c \tag{5}$$

Equation (4) then simplifies to

$$K_a = \frac{[H^+]^2}{c}$$

$$\therefore \qquad [H^+] = (K_a c)^{1/2} \tag{6}$$

Step 4 Collect the data, check the units, and calculate

$$K_a = 1.74 \times 10^{-5} \text{ mol dm}^{-3}$$

$$c = 0.100 \text{ mol dm}^{-3}$$

By equation (6) we therefore obtain

$$[H^+] = 1.74 \times 10^{-5} \text{ mol dm}^{-3} \times 0.100 \text{ mol dm}^{-3})^{1/2}$$

$$= 1.32 \times 10^{-3} \text{ mol dm}^{-3}$$

The hydrogen ion concentration in the solution is therefore 1.32×10^{-3} mol dm^{-3}.

Step 5 Review and check the solution

During this calculation it was assumed that $c - [H^+] = c$. Once we get the answer we should, therefore, check to make sure that this assumption is justifiable. The answer shows that $[H^+] = 1.32 \times 10^{-3}$ mol dm^{-3} = 0.00132 mol dm^{-3}. Therefore $[H^+]$ is about 100 times smaller than c (0.100 mol dm^{-3}); the assumption made during the calculation that $c - [H^+] = c$ is therefore justifiable.

Note

As a general rule, you should always check at the end of a calculation to see whether any assumption made during the calculation is justifiable.

Suppose that in a similar calculation the answer obtained for a quantity x shows that an approximation made (say, $0.100 - x = 0.100$) is not valid. The calculation would then be in error. What can we do then? We should then apply the *method of successive approximations* (read note after Example 7.7). We should substitute the approximate value of x obtained and calculate $(0.100 - x)$; this value for $(0.100 - x)$ should be inserted in the appropriate equation (e.g. equation 4 in the example above) in order to calculate a more accurate value for x. This procedure should be repeated until a self-consistent value for x is obtained as explained in the note after Example 7.7.

In the calculation given the H^+ ions produced by the dissociation of water was ignored; this was because H^+ from water is negligibly small compared with H^+ produced by the dissociation of the weak acid.

Example 8.4 The dissociation of a dibasic acid

Estimate the molar concentrations of H_2S, H^+, HS^- and S^{2-} in

(a) a 0.01 M solution of H_2S in water
(b) a 0.01 M solution of H_2S in 1 M HCl.

The first and second dissociation constants of H_2S are 1×10^{-7} mol dm^{-3} and 1×10^{-14} mol dm^{-3} respectively and $K_w = 1 \times 10^{-14}$ mol^2 dm^{-6}.

Solution to Part (a)

Step 1 Clarify and define the problem

We have to calculate the molar concentrations of H_2S, H^+, HS^-, and S^{2-} present in an aqueous solution of H_2S. To be able to do this we should, as a general rule, first write down *all* the equilibria that exist in the solution.

Three equilibria exist in an aqueous solution of H_2S. Two of these involve the dissociation of the solute (H_2S) and the third involves the dissociation of the solvent (H_2O). The equations for these dissociations are:

$$H_2S(aq) \rightleftharpoons H^+(aq) + HS^-(aq) \tag{1}$$

$$HS^-(aq) \rightleftharpoons H^+(aq) + S^{2-}(aq) \tag{2}$$

$$H_2O(l) \rightleftharpoons H^+(aq) + OH^-(aq) \tag{3}$$

The equilibria existing in the solution are indicated by (1), (2) and (3). How can we calculate the concentrations of the species involved in these equilibria? To calculate these concentrations it is necessary, as a general rule, to make use of the following two important general principles:

(a) that these concentrations are related to one another by the equations for the equilibrium constants
(b) that the concentrations are related to one another by the balanced equations for the reactions.

Even after making use of the principles given above, the calculations are difficult unless certain approximations are made.

Step 2 Select the key equations

The data provide guidance on the selection of the key equations.

Compare the values given in the data for the first dissociation constant (K_1) of H_2S (this corresponds to equilibrium 1), for the second dissociation constant (K_2) of H_2S (this corresponds to equilibrium 2), and for K_w (this corresponds to equilibrium 3). It would then be seen that K_1 is ten million times larger than either K_2 or K_w. This implies that dissociations (2) and (3) are negligibly small when compared with dissociation (1). The concentrations of H^+ and HS^- in the solution may therefore be assumed to be equal to that produced by dissociation (1) alone. *To calculate* $[H^+]$ *and* $[HS^-]$ *in the solution we therefore have to consider only equilibrium (1)*. This equilibrium also involves H_2S. As key equation for the calculation of $[H^+]$, $[HS^-]$, and $[H_2S]$ we therefore select the equation for the equilibrium constant (K_1) of equilibrium (1) which is

$$K_1 = \frac{[H^+][HS^-]}{[H_2S]} \tag{4}$$

We are also asked to calculate $[S^{2-}]$. It is seen that S^{2-} ions appear only in equilibrium (2); to calculate its concentration we therefore have to make use of the equation for the equilibrium constant (K_2) of equilibrium (2) which is

$$K_2 = \frac{[H^+][S^{2-}]}{[HS^-]} \tag{5}$$

Step 3   Derive the equations for the calculations

We have already explained that to calculate $[H^+]$, $[HS^-]$, and $[H_2S]$ we need to consider only equilibrium (1); the calculation would therefore be very similar to that in Example 8.3.

The balanced equation for this equilibrium $(H_2S(aq) \rightleftharpoons H^+(aq) + HS^-(aq))$ shows that

$$[H^+] = [HS^-] \tag{6}$$

and

$$[H_2S] = c - [H^+] \tag{7}$$

where c is 'initial' concentration of H_2S (i.e. the concentration of H_2S before any dissociation takes place). On substituting equations (6) and (7) in equation (4) one obtains

$$K_1 = \frac{[H^+]^2}{c - [H^+]} \tag{8}$$

Equation (8) can be used for the calculation of $[H^+]$ since this is the only unknown.

Once $[H^+]$ is known, $[HS^-]$ and $[H_2S]$ too could be calculated by equations (6) and (7). Then we could calculate $[S^{2-}]$ by equation (5).

Step 4   Collect the data, check the units, and calculate

It is given that

$$K_1 = 1 \times 10^{-7} \text{ mol dm}^{-3}$$

$$K_2 = 1 \times 10^{-14} \text{ mol dm}^{-3}$$

$$c = 1 \times 10^{-2} \text{ mol dm}^{-3}$$

By making use of equation (8) we can therefore calculate $[H^+]$. As explained in Example 8.3 we can, however, make the approximation that

$$c - [H^+] = c$$

Then by equation (8)

$$[H^+] = (K_1 c)^{1/2}$$

$$= (1 \times 10^{-7} \text{ mol dm}^{-3} \times 1 \times 10^{-2} \text{ mol dm}^{-3})^{1/2}$$

$$= 3 \times 10^{-5} \text{ mol dm}^{-3}$$

Now we can calculate $[HS^-]$ by equation (6) and $[H_2S]$ by equation (7). We then find that:

$$[HS^-] = 3 \times 10^{-5} \text{ mol dm}^{-3}$$

$$[H_2S] = 1 \times 10^{-2} \text{ mol dm}^{-3} - 3 \times 10^{-5} \text{ mol dm}^{-3}$$

$$= 1 \times 10^{-2} \text{ mol dm}^{-3}$$

By equation (5), $[S^{2-}]$ is given by

$$[S^{2-}] = \frac{K_2[HS^-]}{[H^+]}$$

$$= 1 \times 10^{-14} \, mol \, dm^{-3}$$

The concentrations of H_2S, H^+, HS^-, and S^{2-} in a $1 \times 10^{-2} \, mol \, dm^{-3}$ aqueous solution of H_2S are therefore $1 \times 10^{-2} \, mol \, dm^{-3}$, $3 \times 10^{-5} \, mol \, dm^{-3}$, $3 \times 10^{-5} \, mol \, dm^{-3}$, and $1 \times 10^{-14} \, mol \, dm^{-3}$ respectively.

Step 5 Review and check the solution

Certain approximations were made during the calculation. Check and make sure that all the assumptions made are justifiable.

Solution to Part (b)

The principles involved and the method of calculation are exactly similar to those given in Part (a). Only a brief outline of the solution is therefore given.

The only difference from the calculation given in Part (a) arises because of the presence initially of HCl in the solution.

Step 1 Clarify and define the problem

In addition to the three dissociations (1), (2), and (3) considered in Part (a), there would be the following dissociation of HCl

$$HCl(aq) \rightarrow H^+(aq) + Cl^-(aq)$$

Since HCl dissociates fully in aqueous solution it follows that the hydrogen ion concentration, produced by the dissociation of 1 M HCl would be $1 \, mol \, dm^{-3}$. Since K_1, K_2, and K_w values are very small it follows that the hydrogen ion concentrations produced by dissociations (1), (2), and (3) are negligibly small (see also answer for $[H^+]$ obtained in Part (a)) compared with $1 \, mol \, dm^{-3}$. We may therefore assume that the hydrogen ion concentration in the solution, $[H^+]$, is due only to the HCl.

$$\therefore \qquad\qquad [H^+] = 1 \, mol \, dm^{-3}$$

Step 2 Select the key equations

For the same reasons given in Part (a), we need to consider only equilibrium (1) for the calculation of $[HS^-]$ and $[H_2S]$, and have to consider equilibrium (3) for the calculation of $[S^{2-}]$.

Since $[H^+]$ in the solution is $1 \, mol \, dm^{-3}$ it follows that the equations (4) and (5) given in Part (a) can be written respectively as

$$K_1 = \frac{(1 \, mol \, dm^{-3}) \, [HS^-]}{[H_2S]} \qquad\qquad (9)$$

$$K_2 = \frac{(1 \, mol \, dm^{-3}) \, [S^{2-}]}{[HS^-]} \qquad\qquad (10)$$

Step 3 Derive the equations for the calculations

We have to calculate two unknowns $[HS^-]$ and $[H_2S]$—by making use of equation (9). To do so we must have another equation relating $[HS^-]$ and $[H_2S]$. Such an equation is obtained from the balanced equation for the dissociation ($H_2S(aq) =$ $H^+(aq) + HS^-(aq)$) which shows that

$$[H_2S] = c - [HS^-] \tag{11}$$

Equation (9) can therefore be rewritten as

$$K_1 = \frac{(1 \text{ mol dm}^{-3})\,[HS^-]}{c - [HS^-]} \tag{12}$$

Equation (12) can be used to calculate $[HS^-]$ since this is the only unknown.

Once $[HS^-]$ is known, equations (10) and (11) can be used to calculate $[S^{2-}]$ and $[H_2S]$ respectively.

Step 4 Collect the data, check the units, and calculate
From data

$$K_1 = 1 \times 10^{-7} \text{ mol dm}^{-3}$$

$$K_2 = 1 \times 10^{-14} \text{ mol dm}^{-3}$$

$$c = 1 \times 10^{-2} \text{ mol dm}^{-3}$$

It was also shown that

$$[H^+] = 1 \text{ mol dm}^{-3}$$

We can now calculate $[HS^-]$ by the application of equation (12). As explained in Part (a) we can assume that $c - [HS^-] = c$ and therefore equation (12) can be written as

$$[HS^-] = \frac{K_1 \times c}{(1 \text{ mol dm}^{-3})}$$

$$= \frac{(1 \times 10^{-7} \text{ mol dm}^{-3}) \times (1 \times 10^{-2} \text{ mol dm}^{-3})}{(1 \text{ mol dm}^{-3})}$$

$$= 1 \times 10^{-9} \text{ mol dm}^{-3}$$

Now we can calculate $[H_2S]$ by equation (11) and $[S^{2-}]$ by equation (10). We then obtain

$$[H_2S] = 1 \times 10^{-2} \text{ mol dm}^{-3}$$

$$[S^{2-}] = \frac{K_2[HS^-]}{1.0 \text{ mol dm}^{-3}}$$

$$= 1 \times 10^{-23} \text{ mol dm}^{-3}$$

The concentrations of H_2S, H^+, HS^-, and S^{2-} in a 1×10^{-2} mol dm^{-3} aqueous solution of H_2S in 1 M HCl are therefore 1×10^{-2} mol dm^{-3}, 1 mol dm^{-3}, 1×10^{-9} mol dm^{-3}, and 1×10^{-23} mol dm^{-3} respectively.

Step 5 Review and check the solution

Check and convince yourself that all the assumptions made during the calculation are justifiable. Recognize also that the calculations need not be done accurately (see page 8) because the data given are not accurate—they are given only to 1 significant figure.

Recognize that each species has only *one* value for its concentration in any particular solution. To explain this statement consider equilibria (1) and (2) and the corresponding equations for K (equations 4 and 5), in Example 8.4(a). The value for $[H^+]$ that should be inserted in equation (4) is the same as that to be inserted in equation (5). (Some students mistakenly believe, for example, that the value for $[H^+]$ that should be inserted in equation (5) is the concentration of H^+ ions produced by dissociation (2) *alone*.)

Comparison of the answers obtained in parts (a) and (b) shows that $[HS^-]$ and $[S^{2-}]$ are much less in the presence of HCl. This is reasonable—it is to be expected both from the equations for the dissociation constants and from Le Chatelier's principle.

Exercise 8.3

From a concentrated aqueous solution of hydrochloric acid of concentration 11.0 mol dm^{-3} how would you prepare a solution having a pH 0?

Exercise 8.4

The pH of a 0.0482 M aqueous solution of a monoprotic acid is 4.41 at 25 °C. Calculate the dissociation constant of the acid.

Exercise 8.5

The dissociation constant of ethanoic acid in water is 1.76×10^{-5} mol dm^{-3} at 25 °C. What would the molar concentration of an aqueous solution of this acid be if its pH is to be 2.00?

Exercise 8.6

A 5.0×10^{-3} M aqueous solution of a weak monoprotic acid has a pH 5.0. Calculate the degree of dissociation of the acid in this solution.

Exercise 8.7

Calculate the pH of a 0.150 M solution of KHSO$_4$. The dissociation constant of the HSO$_4^-$ ion (which is an acid) is 1.26×10^{-2} mol dm^{-3}.

Exercise 8.8

Calculate the concentrations of the various ions in a 0.150 M solution of sulphuric acid. The first dissociation of sulphuric acid is complete; the second dissociation constant is 1.26×10^{-2} mol dm^{-3}.

Exercise 8.9

Calculate the concentrations of the various species in a 0.100 M aqueous solution of H$_3$PO$_4$. For this acid $K_1 = 7.5 \times 10^{-3}$ mol dm^{-3}, $K_2 = 6.2 \times 10^{-8}$ mol dm^{-3}, $K_3 = 1 \times 10^{-12}$ mol dm^{-3}.

Exercise 8.10

In an aqueous solution of the weak base ammonia the following equilibrium exists

$$NH_3(aq) + H_2O(l) \rightleftharpoons NH_4^+(aq) + OH^-(aq)$$

The equilibrium constant for the above reaction is 1.81×10^{-5} mol dm^{-3} at 25 °C. Calculate the dissociation constant for the dissociation of the ammonium ion

$$NH_4^+(aq) \rightleftharpoons NH_3(aq) + H^+(aq)$$

8.3 Aqueous solutions of salts

All salts may be considered to dissociate completely in aqueous solution. The ions formed may disturb the following equilibrium existing due to the dissociation of water

$$H_2O(l) \rightleftharpoons H^+(aq) + OH^-(aq) \tag{1}$$

If a salt does not influence this equilibrium (e.g. any salt of a strong acid and a strong base) the hydrogen ion concentration will be the same as that existing in pure water (1.00×10^{-7} mol dm^{-3} at 25 °C); the solution will then be neutral. If a salt affects the equilibrium existing in water (e.g. any salt of a weak acid or a weak base), the hydrogen ion concentration will change from 1.00×10^{-7} mol dm^{-3} to some other value.

How does a salt affect the equilibrium (1)? This depends on the nature of the ions present in the salt. In a salt MA formed from a weak acid (e.g. CH_3COOH) and a strong base (e.g. KOH) the anion of the salt A$^-$ (e.g. CH_3COO^-) would react with the H$^+$ ions formed from the solvent H_2O to form undissociated acid molecules according to the equation

$$A^-(aq) + H^+(aq) \rightleftharpoons HA(aq) \tag{2}$$

H$^+$ ions, formed by the dissociation of H_2O, are partly removed by reaction (2); the hydrogen ion concentration in the solution will then be *less* than that in pure water. The solution would therefore be alkaline. The removal of H$^+$ from the solution by reaction (2) would lead to further dissociation of H_2O

$$H_2O(l) \rightleftharpoons H^+(aq) + OH^-(aq) \tag{3}$$

Addition of equations (2) and (3) gives the equation for the total reaction occurring which is

$$A^-(aq) + H_2O(l) \rightleftharpoons HA(aq) + OH^-(aq) \tag{4}$$

Reaction (4) can be considered to be a *hydrolysis* reaction, and its equilibrium constant is often known as the hydrolysis constant, K_h. That is

$$K_h = \frac{[HA][OH^-]}{[A^-]} \tag{5}$$

Certain salts will increase the hydrogen ion concentration of a solution by introducing extra H$^+$ ions. For example, in an aqueous solution of an ammonium salt the NH_4^+ formed behaves as an acid since it dissociates according to the equation

$$NH_4^+(aq) \rightleftharpoons NH_3(aq) + H^+(aq)$$

Extra H$^+$ ions are then introduced into the solution; the solution will therefore be acidic.

Example 8.5   Concentrations of the species present in a salt solution

Calculate the concentration of each species present in a 0.050 M aqueous solution of ammonium chloride at 298 K. The dissociation constant of the NH_4^+ ion is 5.6×10^{-10} mol dm^{-3} and K_w is 1.0×10^{-14} mol^2 dm^{-6}.

Solution

Step 1 Clarify and define the problem

To calculate the concentrations of the various species present in a solution, it is first necessary to identify the various species present. For this purpose we should write down *all* the equilibria that exist in the solution.

The two components that go to form the solution are the salt NH_4Cl and the solvent H_2O. The relevant dissociation equilibria existing are

$$H_2O(l) \rightleftharpoons H^+(aq) + OH^-(aq) \qquad (1)$$

$$NH_4Cl \rightarrow NH_4^+(aq) + Cl^-(aq) \qquad \text{(fully dissociated) (2)}$$

$$NH_4^+(aq) \rightleftharpoons NH_3(aq) + H^+(aq) \qquad (3)$$

It is seen that six species are present in the solution; these are:

$$H_2O, H^+(aq), OH^-(aq), NH_4^+(aq), Cl^-(aq), NH_3(aq)$$

(NH_4Cl does not exist since it is fully dissociated. As a general rule assume, unless otherwise specified, that all simple salts are fully dissociated in aqueous solution.)

The concentrations of two of these species—$Cl^-(aq)$ and H_2O—are easily obtained and we shall therefore calculate them here.

$Cl^-(aq)$ occurs only in equation (2); its concentration is equal to the initial concentration of ammonium chloride because the salt is fully dissociated. That is

$$[Cl^-] = 0.050 \text{ mol dm}^{-3}$$

H_2O occurs only in equation (1); the amount of H_2O dissociated according to this equation is negligibly small compared with the amount of H_2O present (K_w is seen to be very small). The concentration of H_2O in the solution can therefore be considered to be the same as that existing in pure water which is 55 mol dm^{-3} at 298 K (see Exercise 8.11). That is

$$[H_2O] = 55 \text{ mol dm}^{-3}$$

Step 2 Select the key equation

We have still to calculate the molar concentrations of the other four species—$H^+(aq)$, $OH^-(aq)$, $NH_4^+(aq)$, and $NH_3(aq)$—present in the solution. How does one select the key equations for these calculations? Guidance in this selection is provided by the given data.

Consider first H^+ ions. A look at all the equilibria existing (see equations 1, 2, and 3) shows that the hydrogen ion concentration in the solution $[H^+]$ is equal to the sum of the hydrogen ion concentrations produced by dissociations (1) and (3). Which of these dissociations is more important? To answer this question look at the equilibrium constant values for these two dissociations. It is then seen that K_a for dissociation (3) is over 1000 times larger than K_w for dissociation (1). Only dissociation (3) need therefore be considered. As key equation we therefore select the equation for the equilibrium constant for the dissociation (3) ($NH_4^+(aq) \rightleftharpoons NH_3(aq) + H^+(aq)$) which is

$$K_a = \frac{[NH_3][H^+]}{[NH_4^+]}$$ (4) **Equation 7.1**

Equation (4) is seen to involve the concentrations of $NH_3(aq)$, $H^+(aq)$, and $NH_4^+(aq)$. It cannot therefore be used to calculate the concentration of OH^- ions.

To calculate the concentration of OH^- ions we must evidently start with an equation which involves OH^- ions. A look at the equations for the dissociation equilibria shows that OH^- ions are involved only in equilibrium (1)

$$H_2O(l) \rightleftharpoons H^+(aq) + OH^-(aq)$$

To calculate $[OH^-]$ we therefore start with the equilibrium constant (ionic product, K_w) for the above dissociation which is

$$K_w = [H^+][OH^-]$$ (5)

Step 3 Derive the equation for the calculations

Let us first see how equation (4) can be used to derive equations for the calculation of $[NH_3]$, $[H^+]$, and $[NH_4^+]$. In this equation only K_a is known, the three concentrations can, however, be calculated using this equation because these are related to the 'initial' concentration of $NH_4^+(aq)$ (which is known) and to each other by the balanced equation for the dissociation reaction.

The balanced equation for the dissociation of $NH_4^+(aq)$ ions

$$NH_4^+(aq) \rightleftharpoons NH_3(aq) + H^+(aq)$$

shows that the following relationships are true

$$[NH_3] = [H^+]$$ (6)

$$[NH_4^+] = c - [H^+]$$ (7)

where c = the 'initial' concentration of NH_4^+ ions.
On substituting equations (6) and (7) in equation (4) we obtain

$$K_a = \frac{[H^+]^2}{c - [H^+]}$$ (8)

Equation (8) can be used to calculate $[H^+]$ since it is the only unknown in the equation.

Once $[H^+]$ is known we can calculate $[NH_3]$, $[NH_4^+]$, and $[OH^-]$ by equations (6), (7), and (5) respectively.

Step 4 Collect the data, check the units, and calculate

$$K_a = 5.6 \times 10^{-10} \text{ mol dm}^{-3}$$

$$K_w = 1.0 \times 10^{-14} \text{ mol}^2 \text{ dm}^{-6}$$

$$c = 0.050 \text{ M} = 0.050 \text{ mol dm}^{-3}$$

$[H^+]$ can be calculated by equation (8). It is, however, possible to simplify this equation before calculation. Since K_a is small, it follows that $[H^+]$ would be small compared with c. We may therefore approximate and write $c - [H^+] = c$. Equation (8) then becomes

$$K_a = \frac{[H^+]^2}{c}$$

$\therefore$

$$[H^+] = (K_a c)^{1/2}$$

$$= (5.6 \times 10^{-10} \text{ mol dm}^{-3} \times 0.050 \text{ mol dm}^{-3})^{1/2}$$

$$= 5.3 \times 10^{-6} \text{ mol dm}^{-3}$$

By equation (6) we then obtain

$$[NH_3] = 5.3 \times 10^{-6} \text{ mol dm}^{-3}$$

By equation (7)

$$[NH_4^+] = c - [H^+]$$

$$= 0.050 \text{ mol dm}^{-3} - 5.3 \times 10^{-6} \text{ mol dm}^{-3}$$

$$= 0.050 \text{ mol dm}^{-3}$$

By equation (5)

$$[OH^-] = \frac{K_w}{[H^+]}$$

$$= \frac{1.0 \times 10^{-14} \text{ mol}^2 \text{ dm}^{-6}}{5.3 \times 10^{-6} \text{ mol dm}^{-3}}$$

$$= 1.9 \times 10^{-9} \text{ mol dm}^{-3}$$

The concentrations of $Cl^-(aq)$, H_2O, $H^+(aq)$, $NH_3(aq)$, $NH_4^+(aq)$, and $OH^-(aq)$ in a 0.050 mol dm^{-3} aqueous solution of ammonium chloride are therefore respectively 0.050 mol dm^{-3}, 55 mol dm^{-3}, 5.3×10^{-6} mol dm^{-3}, 5.3×10^{-6} mol dm^{-3}, 0.050 mol dm^{-3}, and 1.9×10^{-9} mol dm^{-3}.

Step 5 Review and check the solution

Two approximations were made during the calculation:

(a) that $[H^+]$ produced by dissociation (1) is negligibly small compared with that produced by dissociation (3). We must therefore check to see whether this approximation is valid. $[H^+]$ produced by dissociation (1) is seen from the balanced equation to be equal to $[OH^-]$ and is therefore 1.9×10^{-9} mol dm^{-3}. $[H^+]$ produced by dissociation (3) was calculated to be 5.3×10^{-6} mol dm^{-3}. This is $(5.3 \times 10^{-6}/1.9 \times 10^{-9})$ times ($= 300$ times) larger than 1.9×10^{-9} mol dm^{-3}. The approximation made is therefore valid.

(b) that $c - [H^+] = c$. The answer obtained for $[H^+]$ shows that this approximation too is valid.

Example 8.6 Calculation of the hydrogen ion concentration in a salt solution

Calculate the hydrogen ion concentration in a 0.050 M aqueous solution of sodium ethanoate at 298 K. At this temperature the dissociation constant of ethanoic acid is 1.8×10^{-5} mol dm^{-3} and the ionic product of water is 1.0×10^{-14} mol^2 dm^{-6}.

Solution

Step 1 Clarify and define the problem

To calculate the concentration of any species present in a solution we must first write down all the equilibria that exist in the solution. In the solution considered there are two components initially: the solvent H_2O and the salt CH_3COONa. The dissociations and equilibria are then

$$H_2O(l) \rightleftharpoons H^+(aq) + OH^-(aq) \tag{1}$$

$$CH_3COONa \rightarrow CH_3COO^-(aq) + Na^+(aq) \quad \text{fully dissociated} \tag{2}$$

In addition there would be another equilibrium; this is due to the CH_3COO^- ions formed (by equation 2) reacting partially with the solvent H_2O to form CH_3COOH molecules according to the equation

$$CH_3COO^-(aq) + H_2O(l) \rightleftharpoons CH_3COOH(aq) + OH^-(aq) \tag{3}$$

Step 2 Select the key equation

Our problem is to calculate the hydrogen ion concentration in the solution. H$^+$ ions arise *only* from the solvent H_2O, by its dissociation. To calculate [H$^+$] we have, therefore, to focus our attention on the dissociation of H_2O (equation 1). For this dissociation $K_w = [H^+][OH^-]$ and therefore

$$[H^+] = \frac{K_w}{[OH^-]} \tag{4}$$

Step 3 Derive the equation for the calculation

K_w is given in the data; to calculate [H$^+$] by equation (4) we must therefore first calculate [OH$^-$] in the solution.

Hydroxide ions are seen to be produced in equations (1) and (3). Which of these reactions is more important for the production of OH$^-$ ions? To answer this question we must compare the two equilibrium constant values. The equilibrium constant (K_w) for dissociation (1) is given in the data. For equilibrium (3) the K value is not given. Can we calculate it from the data given? Yes, as follows:

For equilibrium reaction (3), since [H$_2$O] is a constant, the equilibrium constant K is given by

$$K = \frac{[CH_3COOH][OH^-]}{[CH_3COO^-]} \tag{5} \text{ Equation 7.1}$$

K can be related to the dissociation constant of ethanoic acid K_a as follows. Since the dissociation of ethanoic acid takes place according to the equation $CH_3COOH(aq) \rightleftharpoons CH_3COO^-(aq) + H^+(aq)$ it follows that

$$K_a = \frac{[CH_3COO^-][H^+]}{[CH_3COOH]}$$

$$\therefore \quad \frac{[CH_3COOH]}{[CH_3COO^-]} = \frac{[H^+]}{K_a} \qquad (6)$$

On replacing $[CH_3COOH]/[CH_3COO^-]$ in equation (5) by equation (6) we obtain

$$K = \frac{[H^+][OH^-]}{K_a}$$

$$= \frac{K_w}{K_a} \qquad (7)$$

Equation (7) can be used to calculate K

$$K = \frac{1.0 \times 10^{-14} \text{ mol}^2 \text{ dm}^{-6}}{1.8 \times 10^{-5} \text{ mol dm}^{-3}}$$

$$= 5.6 \times 10^{-10} \text{ mol dm}^{-3}$$

The K value for the equilibrium (3) (which is 5.6×10^{-10} mol dm^{-3}) is seen to be over 50 000 times larger than the K value for equilibrium (1) (which is 1.0×10^{-14} mol^2 dm^{-6}). The amount of OH$^-$ ions formed by reaction (3) would therefore be very much larger than that formed by dissociation (1). *To calculate [OH$^-$] we therefore need to consider only reaction (3)* for which K is given by equation (5) as

$$K = \frac{[CH_3COOH][OH^-]}{[CH_3COO^-]} \qquad (5)$$

Equation (5) can be used to calculate [OH$^-$] because K and the initial concentration c of CH$_3$COO$^-$ ions are known, and also because the concentrations in equation (5) are related to each other by the balanced equation for the reaction. The balanced equation (equation 3)

$$CH_3COO^-(aq) + H_2O(l) \rightleftharpoons CH_3COOH(aq) + OH^-(aq)$$

shows that

$$[CH_3COOH] = [OH^-]$$

$$[CH_3COO^-] = c - [OH^-]$$

where c is the initial concentration of CH$_3$COO$^-$ ions. Therefore equation (5) can be written as

$$K = \frac{[OH^-]^2}{c - [OH^-]} \qquad (8)$$

Equation (8) can be used to calculate [OH$^-$] since this is the only unknown. It is, however, possible to simplify this equation. Since K is small (5.6×10^{-10} mol dm^{-3}) it follows that [OH$^-$] would be small compared with c. We can therefore make the approximation that $c - [OH^-] = c$. Equation (8) then becomes

$$K = \frac{[OH^-]^2}{c}$$

∴ $[OH^-] = (Kc)^{1/2}$ (9)

We can now obtain the equation required for the calculation by replacing $[OH^-]$ in equation (4) by equation (9). We then obtain

$$[H^+] = \frac{K_w}{(Kc)^{1/2}}$$ (10)

Step 4 Collect the data, check the units, and calculate

$$K_w = 1.0 \times 10^{-14} \text{ mol}^2 \text{ dm}^{-6} \qquad \text{(data)}$$

$$c = 0.050 \text{ mol dm}^{-3} \qquad \text{(data)}$$

$$K = 5.6 \times 10^{-10} \text{ mol dm}^{-3} \qquad \text{(see step 3)}$$

∴ $$[H^+] = \frac{1.0 \times 10^{-14} \text{ mol}^2 \text{ dm}^{-6}}{(5.6 \times 10^{-10} \text{ mol dm}^{-3} \times 0.050 \text{ mol dm}^{-3})^{1/2}}$$

$$= 1.9 \times 10^{-9} \text{ mol dm}^{-3}$$

The required quantity—the hydrogen ion concentration in a 0.050 M aqueous solution of CH_3COONa at 298 K—is therefore 1.9×10^{-9} mol dm^{-3}.

Step 5 Review and check the solution

The answer shows that the solution is alkaline. This is reasonable; by qualitative reasoning we would expect an aqueous solution of CH_3COONa (which is a salt from a weak acid and a strong base) to be alkaline.

Identify all the approximations made during the calculation. Convince yourself that all these approximations are valid.

Exercise 8.11

By making use of **Equation 4.2** ($c = n/V$) show that the concentration of H_2O in water is equal to 55.5 mol dm^{-3} at 298 K. The density of water at 298 K is 1.00×10^3 g dm^{-3} and the molar mass of H_2O is 18.0 g mol^{-1}.

Exercise 8.12

The dissociation constant of the NH_4^+ ion is 5.5×10^{-10} mol dm^{-3} and $K_w = 1.00 \times 10^{-14}$ mol^2 dm^{-6} at 298 K. Calculate (a) the concentrations of the various species in a 0.025 M aqueous solution of $(NH_4)_2SO_4$ (b) the equilibrium constant for the reaction $NH_3(aq) + H_2O(l) \rightleftharpoons NH_4^+(aq) + OH^-(aq)$.

Exercise 8.13

How many moles of sodium ethanoate ($NaOCOCH_3$) must be present in 500 cm^3 of its aqueous solution so that the pH is 8.7? The dissociation constant of ethanoic acid is 1.8×10^{-5} mol dm^{-3} and $K_w = 1.0 \times 10^{-14}$ mol^2 dm^{-6}.

8.4 Buffer solutions

A buffer solution is one which resists changes in the hydrogen ion concentration on the addition of water, acids, or bases.

A buffer solution usually consists either of a mixture of a weak acid and one of its salts, or of a weak base and one of its salts.

The action of a buffer solution is now explained qualitatively with the help of a simple example. Consider a buffer solution containing ethanoic acid (CH_3COOH) and sodium ethanoate (CH_3COONa) in water. The following equilibria would then exist in the solution

$$CH_3COONa(aq) \rightarrow CH_3COO^-(aq) + Na^+(aq)$$
(1) (fully dissociated)
$$CH_3COOH(aq) \rightleftharpoons CH_3COO^-(aq) + H^+(aq) \qquad (2)$$
$$H_2O(l) \rightleftharpoons H^+(aq) + OH^-(aq) \qquad (3)$$

Suppose H^+ ions are now added to the buffer solution. Most of these ions would be 'removed' by reaction with CH_3COO^- ions to form CH_3COOH molecules (by the reverse reaction of equation 2); the hydrogen ion concentration of the solution would therefore not increase significantly.

Suppose OH^- ions (base) are added to a buffer solution. Most of these ions would be 'removed' by reaction with CH_3COOH (acid–base reaction) to form H_2O molecules.

$$OH^-(aq) + CH_3COOH(aq) \rightarrow CH_3COO^-(aq) + H_2O$$

The OH^- concentration, and therefore the H^+ concentration, would therefore not change significantly.

A quantitative understanding of the behaviour of buffer solutions will be obtained after Examples 8.7 and 8.8, and the method for preparing a buffer solution of any desired pH will become clear after Example 8.9.

Example 8.7 The pH of a buffer solution

An aqueous buffer solution at 298 K contains 0.5000 M ethanoic acid and 0.5000 M sodium ethanoate. pK_a for ethanoic acid at 298 K is 4.755, where pK_a is defined by the equation $pK_a = -\log_{10} K_a$. Calculate the pH of the buffer solution.

Solution

Step 1 Clarify and define the problem
We want the pH of a buffer solution of ethanoic acid and sodium ethanoate.

Step 2 Select the key equation
We start with the defining equation for the required quantity—pH; this is given by

$$pH = -\log_{10} [H^+] \qquad \text{(1) } \textbf{Equation 8.1}$$

Step 3 Derive the equation for the calculation
Equation (1) shows that to calculate pH we require the concentration of H^+ ions in the solution. To calculate the concentration of any species in a solution we should, as

a general rule, write down all the equilibria that exist in the solution. In the example considered the equilibria existing are

$$CH_3COONa(aq) \rightarrow CH_3COO^-(aq) + Na^+(aq) \text{ fully dissociated} \quad (2)$$

$$CH_3COOH(aq) \rightleftharpoons CH_3COO^-(aq) + H^+(aq) \quad (3)$$

$$H_2O(l) \rightleftharpoons H^+(aq) + OH^-(aq) \quad (4)$$

It is seen that H^+ ions are produced by dissociations (3) and (4). Which of these dissociations is more important? The equilibrium constant values (K_a and K_w) indicate that dissociation (3) is very much more important than (4). To calculate $[H^+]$ we therefore need to consider only dissociation (3) for which

$$K_a = \frac{[CH_3COO^-][H^+]}{[CH_3COOH]} \qquad \text{Equation 7.1}$$

$$\therefore \qquad [H^+] = \frac{K_a[CH_3COOH]}{[CH_3COO^-]} \qquad (5)$$

On replacing $[H^+]$ in equation (1) by equation (5) we obtain

$$pH = -\log_{10}\left(\frac{K_a[CH_3COOH]}{[CH_3COO^-]}\right)$$

$$= -\log_{10}K_a - \log_{10}\frac{[CH_3COOH]}{[CH_3COO^-]}$$

Since $-\log_{10}K_a$ is defined as pK_a it follows that

$$pH = pK_a + \log_{10}\frac{[CH_3COO^-]}{[CH_3COOH]} \qquad (6)$$

To calculate the pH of the buffer solution using equation (6) we must therefore know $[CH_3COO^-]$ and $[CH_3COOH]$ in the solution.

Consider first the calculation of the concentration of the CH_3COO^- ions. These ions are produced by dissociations (2) and (3). Dissociation (2) takes place completely while dissociation (3) takes place only to a very small extent (particularly because dissociation (3) would be suppressed in the presence of the $CH_3COO^-(aq)$ ions formed by dissociation (1)). $CH_3COO^-(aq)$ ions in the solution may therefore be assumed to be produced only by the dissociation of the salt CH_3COONa. Since CH_3COONa dissociates fully in aqueous solution it is clear that

$$[CH_3COO^-] = [\text{salt}] \qquad (7)$$

where [salt] is used to denote the 'initial' concentration of the salt sodium ethanoate in the solution.

Consider now the calculation of $[CH_3COOH]$. It is seen that CH_3COOH appears only in dissociation (3). Since dissociation (3) is very small we may assume that

$$[CH_3COOH] = [\text{acid}] \qquad (8)$$

where [acid] is used to denote the 'initial' concentration of ethanoic acid in the solution.

Replacement of $[CH_3COO^-]$ and $[CH_3COOH]$ in equation (6) by equations (7) and (8) respectively then gives

$$pH = pK_a + \log_{10} \frac{[salt]}{[acid]} \qquad (9)$$

Step 4 Collect the data, check the units, and calculate

$$pK_a = 4.755$$

$$[salt] = 0.5000 \text{ mol dm}^{-3}$$

$$[acid] = 0.5000 \text{ mol dm}^{-3}$$

$$\therefore \qquad pH = 4.755 + \log_{10}\left(\frac{0.5000}{0.5000}\right)$$

$$= 4.755$$

Step 5 Review and check the solution

The calculation was based on the assumption that the dissociation of CH_3COOH (equation 2) is negligibly small. This assumption should therefore be checked now to see whether it is valid.

The answer shows that pH $=$ 4.755; this means that the hydrogen ion concentration in the solution is between 10^{-4} mol dm^{-3} and 10^{-5} mol dm^{-3}. The hydrogen ions in the solution are produced, as explained earlier, mainly by the dissociation of CH_3COOH. The balanced equation for this dissociation shows that if $[H^+] = 5 \times 10^{-4}$ mol dm^{-3}, then $[CH_3COOH] = (0.5000 - 5 \times 10^{-4})$ mol dm$^{-3} =$ 0.5000 mol dm^{-3} (since 5×10^{-4} is negligibly small compared with 0.5000). The approximation made during the calculation that the dissociation of CH_3COOH is negligibly small is therefore valid.

Note

Most calculations involving buffer solutions can be done by making use of equation (7).

$$pH = pK_a + \log_{10} \frac{[salt]}{[acid]} \qquad (7)$$

It may therefore be worthwhile remembering this equation but we do not include it as one of the defining equations. It can be derived by applying **Equations 7.1** and **8.1** to each particular system. In Examples 8.8 and 8.9 this equation is used without working through the complete derivation.

Example 8.8 pH changes in a buffer solution

Consider the buffer solution given in Example 8.7. Calculate the pH change if to

100.0 cm^3 of the buffer solution is added

(a) 10.00 cm^3 of water
(b) 10.00 cm^3 of a 0.1000 M HCl solution
(c) 10.00 cm^3 of a 0.1000 M NaOH solution.

Solution

Step 1   Clarify and define the problem

We want to calculate the pH change in a buffer solution on the addition of another solution. Evidently

pH change = pH of the final solution − pH of the initial solution

The pH of the initial buffer solution is known (pH = 4.755; calculated in Example 8.7). To calculate the pH change we have therefore to calculate only the pH of the final buffer solution—i.e. of the solution after the additions indicated in the problem.

Steps 2 and 3   Key equation and the equation for the calculation

The solutions obtained after the addition of water, HCl, or NaOH are also buffer solutions of ethanoic acid and sodium ethanoate. The equation for calculation, and its derivation is therefore the same as that given in Example 8.7 and is

$$pH = pK_a + \log_{10} \frac{[\text{salt}]}{[\text{acid}]} \qquad (1)$$

Step 4   Collect the data, check the units, and calculate

pK_a is given in the data (4.755). To calculate pH by equation (1) we therefore require [salt] and [acid] in the final solution.

Part (a)

When water is added to the buffer solution the only change taking place is dilution. Since 10.00 cm^3 of water is added to 100.0 cm^3 of the buffer solution the final volume of the solution would be 110.0 cm^3 (we assume here that volumes are additive; the error involved in this assumption is small with dilute aqueous solutions). By the application of **Equation 4.2** ($c = n/V$) it is then easy to see that

$$[\text{salt}] = (0.5000 \text{ mol dm}^{-3}) \times \frac{100.0 \text{ cm}^3}{110.0 \text{ cm}^3}$$

$$= 0.4545 \text{ mol dm}^{-3}$$

Similarly

$$[\text{acid}] = 0.4545 \text{ mol dm}^{-3}$$

The pH of the buffer solution after the addition of water is therefore given by equation (1) as

$$pH = 4.755 + \log_{10} \frac{0.4545}{0.4545}$$

$$= 4.755$$

The pH of the buffer solution, therefore, does not change on the addition of water.

Part (b)

Equation (1) shows that the pH of a buffer solution depends on [salt] and [acid]. When HCl is added, the H^+ ions from it would react with CH_3COO^- ions in the buffer solution to form CH_3COOH. The concentration of CH_3COO^- (i.e. [salt]) would therefore decrease while the concentration of CH_3COOH (i.e. [acid]) would increase.

The changes in the concentration of the acid and the salt could be calculated by making use of the balanced equation for the reaction. The calculation is tabulated in Table 8.1.

Table 8.1   Calculation for Example 8.8

Balanced equation	CH_3COO^-	+	HCl	→	$CH_3COOH + Cl^-$
moles in 100 cm³ buffer solution	0.0500				0.0500
moles in 10.00 cm³ 0.1000 M HCl			0.0010		
moles n present after addition of HCl to buffer solution	$(0.0500 - 0.0010)$ $= 0.0490$				$(0.0500 + 0.0010)$ $= 0.0510$
* Concentration c/mol dm⁻³	$\left(\dfrac{0.0490}{0.110}\right)$ $= 0.4455$				$\left(\dfrac{0.0510}{0.110}\right)$ $= 0.4636$

* $c = \dfrac{n}{V}$ and $V = 110 \text{ cm}^3 = 0.110 \text{ dm}^3$

The concentrations of CH_3COO^- (i.e. salt) and CH_3COOH (i.e. acid) in the final solution are therefore 0.4455 mol dm⁻³ and 0.4636 mol dm⁻³ respectively. The pH of the buffer solution after the addition of HCl could now be calculated by equation (1).

$$pH = pK_a + \log_{10} \frac{[\text{salt}]}{[\text{acid}]}$$

$$= 4.755 + \log_{10} \left(\frac{0.4455}{0.4636}\right)$$

$$= 4.738$$

The pH of the initial buffer solution is 4.755.

∴   pH *decrease* $= 4.755 - 4.738 = 0.017$

The pH of 100.0 cm³ of the buffer solution therefore decreases by 0.017 when 10.00 cm³ of 0.1000 M HCl are added to it.

Part (c)

When NaOH is added the following reaction takes place

$$NaOH(aq) + CH_3COOH(aq) \rightarrow CH_3COONa(aq) + H_2O(l)$$

The concentration of salt (CH_3COONa) would therefore increase while the concentration of acid (CH_3COOH) would decrease. A calculation, similar to that given under (b), would show that

$$[salt] = 0.4636 \text{ mol dm}^{-3}$$

$$[acid] = 0.4455 \text{ mol dm}^{-3}$$

$$\therefore \qquad pH = 4 \cdot 755 + \log_{10} \frac{0.4636}{0.4455} = 4.772$$

$$\therefore \qquad pH \text{ increase} = 4.772 - 4.755 = 0.017$$

The pH of 100.0 cm³ of the buffer solution therefore increases by 0.017 when 10.00 cm³ of 0.1000 M NaOH are added to it.

Note

Study carefully the above example; it illustrates *quantitatively* the principles involved in buffer action. The pH changes are seen to be small on the addition of 10 cm³ of 0.1 M HCl or NaOH to 100 cm³ of buffer. If 10 cm³ of 0.1 M HCl are added to 100 cm³ of water the pH change would be large (4.96)—do this calculation as an exercise.

Example 8.9 Preparation of a buffer solution

Suppose the following weak acids are available in the laboratory: CH_3COOH ($pK = 4.76$), $ClCH_2COOH$ ($pK = 2.86$), CCl_3COOH ($pK = 0.70$), and $H_2C_2O_4$ ($pK_1 = 1.23$, $pK_2 = 4.19$). A 1.00 M solution of sodium hydroxide is also available. Which acid would you select, and how would you prepare a buffer solution of pH 3.00 at 298 K? All data are given for 298 K.

Solution

Step 1 Clarify and define the problem

We have to prepare a buffer solution of a particular pH. To see how this can be done, we have first to derive an equation which shows how the pH of a buffer solution depends on concentrations and K_a values and then make use of this equation to select the conditions which would give the required pH.

Steps 2 and 3 Key equation and equation for the calculation

The equation for the calculation, and its derivation, is the same as that given in Example 8.7 and is

$$pH = pK_a + \log_{10} \frac{[salt]}{[acid]} \qquad (1)$$

This equation suggests the method that should be used for preparing a buffer solution of any desired pH. For any acid (which fixes the pK_a value) we should vary the ratio [salt]/[acid] so as to obtain the desired pH.

It is easy to show (see Exercise 8.17) that the best buffer solution is obtained when [salt]/[acid] = 1, i.e. when pH = pK_a. To prepare the best buffer solution from the acids available, the first step is therefore to select an acid whose pK_a value is closest to the desired pH. $ClCH_2COOH$ ($pK_a = 2.86$) is therefore the best choice here.

The ratio of [salt]/[acid] necessary to prepare a buffer solution of any pH could now be calculated by making use of equation (1) which on rearrangement gives

$$\log_{10} \frac{[\text{salt}]}{[\text{acid}]} = \text{pH} - \text{p}K_a$$

$$\therefore \quad \frac{[\text{salt}]}{[\text{acid}]} = \text{antilog} \ (\text{pH} - \text{p}K_a) \tag{2}$$

The [salt[/[acid] ratio required to prepare a buffer solution of any desired pH can therefore be calculated by equation (2).

Step 4 Collect the data, check the units, and calculate
For the acid selected, $\text{p}K_a = 2.86$. By equation (2), the [salt]/[acid] ratio required to prepare a buffer solution of pH 3.00 is given by

$$\frac{[\text{salt}]}{[\text{acid}]} = \text{antilog} \ (3.00 - 2.86)$$

$$= 1.38$$

To prepare a buffer solution of pH 3.00, you should therefore prepare a solution of $ClCH_2COOH$ of known concentration, and to this add the calculated amount of the sodium hydroxide solution so that in the buffer solution then obtained the ratio $[ClCH_2COONa]/[ClCH_2COOH]$ would be equal to 1.38.

(For example, calculation using the balanced equation for the reaction between $ClCH_2COOH$ and NaOH would show that the addition of 57.9 cm^3 of 1.00 M NaOH to 100 cm^3 of 1.00 M $ClCH_2COOH$ would result in a $[ClCH_2COONa]/[ClCH_2COOH]$ ratio of 1.38—check this calculation as an exercise.)

Step 5 Review and check the solution

Exercise 8.14

10.00 cm^3 of a 0.1000 M sodium hydroxide solution are added to 30.00 cm^3 of a 0.1000 M ethanoic acid solution. Calculate the pH of the buffer solution so obtained. $\text{p}K_a$ is 4.76.

Exercise 8.15

What volume of a 0.1000 M solution of sodium hydroxide must be added to 30.00 cm^3 of a 0.1000 M ethanoic acid solution so as to prepare a buffer solution of pH 4.80? K_a for ethanoic acid is 1.76×10^{-5} mol dm^{-3}.

Exercise 8.16

Each of the ethanoic acid—ethanoate buffer solutions given below has a pH 4.755 since the $\text{p}K_a$ is 4.755

(a) 1.000 M CH_3COOH, 1.000 M CH_3COONa
(b) 0.5000 M CH_3COOH, 0.5000 M CH_3COONa
(c) 0.1000 M CH_3COOH, 0.1000 M CH_3COONa.

Suppose 10.00 cm³ of 0.5000 M HCl is added to 100.0 cm³ of each of the above buffer solutions. Calculate the pH change in each case.

Compare the results. Recognize that for the same ratio of acid to salt, the efficiency of a buffer solution increases with increasing concentration. (*Note*: the *efficiency* of a buffer solution, generally known as the *buffer capacity*, is the capacity of the solution to resist changes in pH on the addition of acids or bases.)

Exercise 8.17

Calculate the pH change when 10.00 cm³ of 0.5000 M HCl is added to 100.0 cm³ of each of the following buffer solutions. pK_a is 4.755.

(a) 0.800 M CH_3COOH, 1.200 M CH_3COONa
(b) 1.000 M CH_3COOH, 1.000 M CH_3COONa
(c) 1.200 M CH_3COOH, 0.800 M CH_3COONa
(d) 1.400 M CH_3COOH, 0.600 M CH_3COONa

Compare the results. Note that for the same total solute concentration (2 M in each case) the efficiency of the buffer solution is maximum when the concentrations of the acid and salt are equal. That is when [salt]/[acid] = 1. Note also from the results that the greater the deviation of the ratio [salt]/[acid] from 1, the less efficient is the buffer solution.

Exercise 8.18

A 0.0500 M aqueous solution of potassium hydrogen phthalate, which is a buffer solution, has a pH of 4.01 at 298 K. Calculate the change in pH if 10.00 cm³ of (a) 0.1000 M HCl solution (b) 0.1000 M NaOH solution, are added to 100.0 cm³ of the buffer solution.

8.5 Solubility equilibria

The equilibria considered so far involved H^+ and OH^- ions. Other types of ionic equilibria are also common; examples are the dissociation of complex ions and of sparingly soluble electrolytes.

Consider an aqueous solution of a sparingly soluble electrolyte such as Ag_2CrO_4. The following equilibrium then exists

$$Ag_2CrO_4(s) \rightleftharpoons 2Ag^+(aq) + CrO_4^{2-}(aq)$$

Since Ag_2CrO_4 is a solid, its concentration is a constant. On applying the law of chemical equilibrium we therefore get

$$K = [Ag^+]^2[CrO_4^{2-}]$$

K is a constant that depends only on the temperature; it is known as the *solubility product*.

Example 8.10 Calculation of solubility from solubility product

The solubility product of Ag_2CrO_4 is 9.0×10^{-12} mol^3 dm^{-9} at 298 K. At this temperature calculate the solubility of Ag_2CrO_4 in (a) water (b) a 0.100 M $AgNO_3$ solution.

Solution to Part (a)

Step 1 Clarify and define the problem

We have to calculate the solubility of Ag_2CrO_4 from its solubility product; to do so we must derive the equation that shows the relation between solubility and solubility product.

Step 2 Select the key equation

The solubility product K of Ag_2CrO_4 is defined by the equation

$$K = [Ag^+]^2[CrO_4^{2-}]$$ (1) **Equation 7.1**

Step 3 Derive the equation for the calculation

Equation (1) can be given in terms of the required quantity—the solubility—as follows. Let x be the solubility (i.e. the molar concentration of the solute in a saturated solution) of the solute (Ag_2CrO_4) in water. Since 1 mole of Ag_2CrO_4 completely dissociates to give 2 moles of Ag^+ ions and 1 mole of CrO_4^{2-} ions it follows that in a saturated solution of Ag_2CrO_4

$$[Ag^+] = 2x \quad \text{and} \quad [CrO_4^{2-}] = x$$

On substituting these values into equation (1) we get

$$K = (2x)^2 \times x = 4x^3$$

$\therefore$
$$x = \left(\frac{K}{4}\right)^{1/3}$$ (2)

Step 4 Collect the data, check the units, and calculate

$$K = 9.0 \times 10^{-12} \, mol^3 \, dm^{-9}$$

$\therefore$
$$x = \left(\frac{9.0 \times 10^{-12} \, mol^3 \, dm^{-9}}{4}\right)^{1/3}$$

$$= 1.31 \times 10^{-4} \, mol \, dm^{-3}$$

The solubility of Ag_2CrO_4 is therefore $1.31 \times 10^{-4} \, mol \, dm^{-3}$.

Step 5 Review and check the solution

Solution to Part (b)

Step 1 Clarify and define the problem

We have to calculate the solubility of Ag_2CrO_4 in a 0.100 M $AgNO_3$ solution given the solubility product.

Step 2 Select the key equation

The solubility product K of Ag_2CrO_4 is defined by the equation

$$K = [Ag^+]^2[CrO_4^{2-}]$$ (1) **Equation 7.1**

Step 3 Derive the equation for the calculation

The concentrations of the ions in equation (1) can be related to the required quantity—the solubility—as follows. Let x be the solubility of the solute (Ag_2CrO_4) in a $AgNO_3$ solution of concentration c. Then it follows that

$$[Ag^+] = [Ag^+] \text{ from } AgNO_3 + [Ag^+] \text{ from solubility of } Ag_2CrO_4$$

$$= c + 2x \text{ mol dm}^{-3} \tag{2}$$

and, $[CrO_4^{2-}] = x \text{ mol dm}^{-3} \tag{3}$

On replacing $[Ag^+]$ and $[CrO_4^{2-}]$ in equation (1) by equations (2) and (3) we obtain

$$K = (c + 2x)^2 \times x$$

Since K is small it follows that x too would be small. We may therefore assume that $(c + 2x) = c$ and therefore rewrite the last equation as

$$K = c^2 \times x$$

$$\therefore \qquad x = \frac{K}{c^2} \tag{4}$$

Step 4 Collect the data, check the units, and calculate

$$K = 9.0 \times 10^{-12} \text{ mol}^3 \text{ dm}^{-9}$$

$$c = 0.100 \text{ mol dm}^{-3}$$

$$\therefore \qquad x = \frac{9.0 \times 10^{-12} \text{ mol}^3 \text{ dm}^{-9}}{(0.100 \text{ mol dm}^{-3})^2}$$

$$= 9.0 \times 10^{-10} \text{ mol dm}^{-3}$$

The solubility of Ag_2CrO_4 in a 0.100 mol dm^{-3} solution of $AgNO_3$ is therefore 9.0×10^{-10} mol dm^{-3}.

Step 5 Review and check the solution

The approximation made during the calculation that $(c + 2x) = c$ is seen to be justified because the answer shows that x is very small compared with c.

The answers appear reasonable since they show that the solubility of Ag_2CrO_4 in 0.100 M Ag^+ ions is much smaller than that in water. This is to be expected from Le Chatelier's principle also. The reduction in solubility due to the presence of a 'common' ion (Ag^+ ion, in this case) is sometimes known as the '*common-ion effect*'.

Example 8.11 Precipitation titration of a chloride solution with silver nitrate

A well known precipitation titration is that of a chloride solution, containing CrO_4^{2-} ions as indicator, with silver nitrate (in the burette). 25.0 cm^3 of a 0.100 M solution of NaCl, also containing 0.01 M CrO_4^{2-} ions, is titrated with 0.100 M $AgNO_3$. Which salt will precipitate first? The solubility products, at room temperature, of AgCl and Ag_2CrO_4 are 1.6×10^{-10} mol^2 dm^{-6} and 9.0×10^{-12} mol^3 dm^{-9} respectively.

Solution

Step 1 Clarify and define the problem

It is necessary to clarify the problem and analyse it carefully. The solution in the titration flask contains 0.100 M Cl^- and 0.01 M CrO_4^{2-}. Silver ions are added from the burette. The following two precipitation reactions are then possible:

$$Cl^-(aq) + Ag^+(aq) \rightarrow AgCl(s)$$

$$CrO_4^{2-}(aq) + 2Ag^+(aq) \rightarrow Ag_2CrO_4(s)$$

We have to find out which of these two salts (whether AgCl or Ag_2CrO_4) precipitates first. To find this we have to calculate the silver ion concentrations that are necessary to precipitate the Cl^- present (as AgCl) and the CrO_4^{2-} present (as Ag_2CrO_4), and then compare them.

Step 2 Select the key equations

The silver ion concentration $[Ag^+]_1$ required to precipitate the Cl^- ions present in the solution could be calculated by making use of the defining equation for the solubility product of AgCl (K_{AgCl}) which is

$$K_{AgCl} = [Ag^+]_1[Cl^-] \qquad \textbf{Equation 7.1}$$

$$\therefore \qquad [Ag^+]_1 = \frac{K_{AgCl}}{[Cl^-]} \qquad (1)$$

Similarly the silver ion concentration, $[Ag^+]_2$, required to precipitate the CrO_4^{2-} ions present in the solution, could be calculated by making use of the defining equation for $K_{Ag_2CrO_4}$ which is

$$K_{Ag_2CrO_4} = [Ag^+]_2^2[CrO_4^{2-}] \qquad \textbf{Equation 7.1}$$

$$\therefore \qquad [Ag^+]_2 = \left(\frac{K_{Ag_2CrO_4}}{[CrO_4^{2-}]}\right)^{1/2} \qquad (2)$$

Step 3 Derive the equation for the calculation

Equations (1) and (2) could be used to calculate $[Ag^+]_1$ and $[Ag^+]_2$.

Step 4 Collect the data, check the units, and calculate

$$K_{AgCl} = 1.6 \times 10^{-10} \, mol^2 \, dm^{-6}$$

$$K_{Ag_2CrO_4} = 9.0 \times 10^{-12} \, mol^3 \, dm^{-9}$$

$$[Cl^-] = 0.100 \, mol \, dm^{-3}$$

$$[CrO_4^{2-}] = 0.01 \, mol \, dm^{-3}$$

By equations (1) and (2) we can now calculate $[Ag^+]_1$ and $[Ag^+]_2$.

$$[Ag^+]_1 = \frac{1.6 \times 10^{-10} \, mol^2 \, dm^{-6}}{0.100 \, mol \, dm^{-3}}$$

$$= 1.6 \times 10^{-9} \, mol \, dm^{-3}$$

$$[Ag^+]_2 = \left\{ \frac{9.0 \times 10^{-12} \text{ mol}^3 \text{ dm}^{-9}}{0.01 \text{ mol dm}^{-3}} \right\}^{1/2}$$

$$= 3.0 \times 10^{-5} \text{ mol dm}^{-3}$$

Compare $[Ag^+]_1$ and $[Ag^+]_2$. It is then seen that $[Ag^+]_1$ is smaller than $[Ag^+]_2$. A smaller silver ion concentration is thus sufficient to precipitate the chloride ions present as AgCl; this salt would therefore precipitate first.

Step 5   Review and check the solution

Exercise 8.19

At 25 °C, 250 cm^3 of a saturated solution of calcium hydroxide contains 2.85×10^{-3} g of the solute. Calculate (a) the solubility product of calcium hydroxide (b) the solubility of calcium hydroxide in the presence of 0.100 M sodium hydroxide.

Exercise 8.20

The solubility product of $CaCrO_4$ can be determined in the laboratory by titrating a *saturated* solution of it with a suitable reducing agent. 25.00 cm^3 of a saturated solution of $CaCrO_4$ was added to excess acidified potassium iodide. The iodine then liberated required 28.55 cm^3 of 0.0200 M $S_2O_3^{2-}$ solution. Calculate the solubility product of $CaCrO_4$.

Exercise 8.21

A 0.100 M $AgNO_3$ solution was added gradually to 50 cm^3 of a solution containing 0.100 M Cl^- ions and 0.100 M I^- ions. Show from solubility product data that AgI will precipitate first. When AgCl just starts to precipitate what will be (a) the moles (b) the mass, of AgI precipitated. Calculate also the concentrations of the various species in the solution when AgCl just starts to precipitate. The solubility products of AgCl and AgI at 25 °C are 1.56×10^{-10} mol^2 dm^{-6} and 1.50×10^{-16} mol^2 dm^{-6} respectively.

Exercise 8.22

Calculate the concentrations of the various species in a 0.1000 M aqueous solution of a weak acid HA given that the concentration of hydrogen ions is 8.2×10^{-3} mol dm^{-3}.

Exercise 8.23

Calculate the concentrations of H^+ and A^{2-} in a 0.1000 M solution of a dibasic acid H_2A if the concentrations of H_2A and HA^- are 0.0604 mol dm^{-3} and 0.0391 mol dm^{-3} respectively.

Exercise 8.24

Calculate the pH of an aqueous 1.0×10^{-4} M HCl solution at 25 °C if it is diluted (a) 10 times (b) 1000 times (c) 10^5 times (d) 10^6 times.

Exercise 8.25

Suppose you want to prepare a sulphuric acid solution of pH = 1.000. What should the concentration of the acid be? Assume that the first dissociation of the acid is complete and that $K_2 = 1.26 \times 10^{-2}$ mol dm^{-3}.

Exercise 8.26

What volume of 0.1000 M HCl must be added to 30.00 cm^3 of a 0.1000 M solution of ammonia so as to prepare a buffer solution of pH 9.40? The dissociation constant of NH$_4^+$ is 5.55×10^{-10} mol dm^{-3}.

Review of Chapter 8

All the examples and exercises in this chapter, like those in Chapter 7, are based on the application of just one important but simple equation, **Equation 7.1** for the equilibrium constant.

The ionic product, dissociation constant, hydrolysis constant and solubility product are all equilibrium constants. Calculations involving all these physical quantities are therefore similar.

Equilibrium calculations involving the dissociation of water or of weak electrolytes in solution are similar to those involving other dissociations e.g. gas dissociation. A buffer solution is merely a weak electrolyte (e.g. CH$_3$COOH) in the presence of one of the products of its dissociation (e.g. CH$_3$COO$^-$ ions). Calculations involving buffer solutions are therefore essentially calculations based on weak electrolyte dissociation equilibria. Since solubility product is also an equilibrium constant, calculations involving solubility products are similar to calculations involving other equilibrium constants.

It is convenient to express small values of concentrations and equilibrium constants using a logarithmic scale, for example

$$pH = -\log_{10}[H^+]$$

$$pK_a = -\log_{10} K_a$$

and, in general, for property x this is represented as

$$pX = -\log_{10} X \qquad\qquad \textbf{8.1}$$

9 | Energy changes

The study of the various types of energy change, known as thermodynamics, provides a general and important method for understanding, interpreting, and correlating the various facts and laws of the physical sciences.

Thermodynamics is based on three laws, known as the first, second, and third laws. In this chapter we consider problems based only on the first law (particularly, heats or enthalpies of reactions).

9.1 Law of conservation of energy and Einstein's equation

The law of conservation of energy states that energy cannot be created or destroyed. That is

total energy before any process = total energy after the process

Einstein showed that mass is related to energy and that the relationship between mass m and energy E is given by the equation

$$E = mc^2 \tag{1}$$

where c is the velocity of light (3.00×10^8 m s^{-1}).

Example 9.1 Energy released in a nuclear reaction

In the sun and the stars (which are 'huge balls' of fire with temperatures greater than 10^6 K) and in the hydrogen bomb, energy is released from *nuclear fusion* reactions. Three moles of deuterium nuclei ($^2_1$H) fuse together to give 1 mole of helium atoms according to the equation

$$3\,^2_1\text{H} \rightarrow\, ^4_2\text{He} +\, ^1_1\text{H} +\, ^1_0\text{n}$$

If the energy released per mole of helium formed is 2.0×10^{12} J, calculate the change in mass during this reaction.

Solution

Step 1 Clarify and define the problem

The change in mass (Δm) is required. It is given by:

$$\Delta m = m_p - m_r \tag{1 (see note 1 below)}$$

where m_p and m_r are the total masses of the products and of the reactants respectively.

We have to calculate the mass change (Δm) from the data given—energy change (ΔE). To do so we have to derive an equation which shows the relationship between Δm and ΔE; the right side of equation (1) must therefore be related to ΔE.

Step 2 Select the key equation

$m_p - m_r$ in equation (1) must be related to ΔE (i.e. to $E_p - E_r$).

This could be done by the application of Einstein's equation; if we apply this equation to the products and to the reactants we obtain the following two equations

$$m_p = \frac{E_p}{c^2} \qquad (2)$$

$$m_r = \frac{E_r}{c^2} \qquad (3)$$

Step 3 Derive the equation for the calculation

On replacing m_p and m_r in equation (1) by equations (2) and (3) we obtain

$$\Delta m = \frac{E_p}{c^2} - \frac{E_r}{c^2} = \frac{E_p - E_r}{c^2}$$

$$= \frac{\Delta E}{c^2} \qquad (4)$$

Step 4 Collect the data, check the units, and calculate

$$\Delta E = \text{energy of products } (E_p)$$

$$- \text{energy of reactants } (E_r)$$

$$= \text{energy } absorbed$$

$$= -(\text{energy } released)$$

$$= -(2.0 \times 10^{12} \text{ J})$$

$$c = 3.0 \times 10^8 \text{ m s}^{-1}.$$

By the application of equation (4) we therefore get

$$\Delta m = \frac{-2.0 \times 10^{12} \text{ J}}{(3.0 \times 10^8 \text{ m s}^{-1})^2}$$

$$= -2.2 \times 10^{-5} \text{ J m}^{-2} \text{ s}^2$$

$$= -2.2 \times 10^{-5} \text{ kg} \qquad \text{(see note 2)}$$

The minus sign for Δm shows that there is a *decrease* in mass; the decrease in mass during the reaction is therefore 2.2×10^{-5} kg.

Step 5 Review and check the solution

The unit in the answer is correct and the number of significant figures is appropriate.

Notes

1 The symbol Δ is used to denote 'change in'. Thus Δm = change in m, ΔE = change in E, and in general Δx = change in x, where x is any quantity.

For any quantity x, the change in x (i.e. Δx) is defined by

$$\Delta x = \text{final value of } x - \text{initial value of } x$$

The change in mass Δm after a reaction would therefore be given by

$$\Delta m = \text{total mass at end of reaction } - \text{total mass before reaction.}$$

$$= \text{total mass of products } (m_p) - \text{total mass of reactants } (m_r).$$

2 The SI unit of energy is the joule (J). The joule is related to the basic SI units by

$$1 \text{ J} = 1 \text{ kg m}^2 \text{ s}^{-2}$$

(see section 2.4, and also the note after Example 5.1)

3 Recognize that a large amount of energy is released for a small decrease in mass. The energy released here (2×10^{12} J) is approximately 1000 million times greater than that released during a chemical reaction.

Exercise 9.1

Study Einstein's equation. Recognize that since c is very large the energy E corresponding to even a small mass is very large.

How many million kilojoules of energy would be associated with a mass of 1.0 g?

Exercise 9.2

The controlled fission of uranium (U) in a nuclear reactor is used in many countries for the generation of electricity. One of the main reactions taking place is

$$^{235}_{92}\text{U} + ^{1}_{0}\text{n} \rightarrow ^{141}_{56}\text{Ba} + ^{92}_{36}\text{Kr} + 3^{1}_{0}\text{n}$$

The decrease in mass during the reaction is 2.1×10^{-4} kg mol^{-1}. Calculate the change in energy (ΔE). Is energy released or absorbed?

9.2 First law of thermodynamics

The first law of thermodynamics is an *extension* of the law of conservation of energy.

Consider any system; let its total energy, which is called the *internal energy*, be denoted by U. Suppose now that an amount of heat energy q, and an amount of mechanical energy (i.e. work) w, are supplied to the system. By the law of conservation of energy, the internal energy change ΔU has to be equal to $q + w$. That is

$$\Delta U = q + w \qquad\qquad \textbf{Equation 9.1}$$

The first law states that the internal energy change ΔU depends *only* on the initial and final states of a system.

Example 9.2 Calculation of change in internal energy

When 1.00 mol of molecules of a non-ideal gas is compressed the work done (which can be calculated since it is equal to the applied pressure multiplied by the change in volume) is 280 J. The heat energy then evolved is 300 J. Calculate the change in internal energy of the gas.

Solution

Step 1 Clarify and define the problem

We have to calculate the change in internal energy, ΔU.

Step 2 Select the key equation

To calculate the change in internal energy ΔU we make use of the equation

$$\Delta U = q + w \qquad\qquad \text{(1) Equation 9.1}$$

Step 3 Derive the equation for the calculation

Equation (1) can be used for the calculation of ΔU.

Step 4 Collect the data, check the units, and calculate

We must be careful about the signs (whether + or −) for q and w in equation (1). Here w is the work done on the system and q is the heat absorbed by the system. w will, therefore, have a + sign if work is done *on* the system (e.g. compression of a gas) and a −ve sign if work is done *by* the system (e.g. expansion of a gas). Similarly q will have a +ve value if heat is absorbed and a negative value if heat is evolved.

In the example considered work is done on the system and heat is evolved. From the data given it therefore follows that

$$w = +280 \text{ J}$$

$$q = -300 \text{ J}$$

$$\therefore \qquad \Delta U = q + w$$

$$= (-300 \text{ J}) + (+280 \text{ J}) = -20 \text{ J}$$

The internal energy of the gas therefore decreases by 20 J.

Step 5 Review and check the solution

Exercise 9.3

Calculate (a) the work done by the vapour formed (b) the change in internal energy, when 1.00 mol of water molecules is vaporized at 373 K at a constant pressure of 1.013×10^5 Pa. The heat absorbed during the process (which is known as the molar enthalpy of vaporization, formerly called latent heat of vaporization) is 40.7 kJ mol^{-1}.

Remember that the work done by a gas when it expands at a constant pressure p is given by $p \times \Delta V$, where ΔV is the change in volume.

Exercise 9.4

In an *ideal* gas there are no forces of attraction or repulsion between the molecules present. The internal energy of the gas will then not change when the gas is either expanded or contracted. If the work done when an ideal gas expands is 280 J, calculate the heat absorbed.

9.3 Heat changes at constant pressure—enthalpy changes

For any process taking place at constant pressure, the heat change is said to be equal to ΔH, where ΔH is the change in a physical property of the system known as enthalpy (H). It is the normal practice, therefore, to denote heat changes at constant pressure by the symbol ΔH.

ΔH (like ΔU) depends *only* on the initial and final states of the system. From this it follows that ΔH values (i.e. the enthalpy changes, or heats of reactions at constant pressure) are additive. The calculations in this chapter involving enthalpy changes of reactions (see Examples 9.3, 9.4, and 9.5) will be seen to be based on this simple principle—the *principle of additivity of enthalpy changes of reactions*.

The enthalpy change in a process is found to depend on the conditions (temperature, pressure, concentration). When each reactant and each product is present in its standard state (see Table 9.1) the enthalpy change is known as the standard enthalpy change and is denoted as ΔH . Enthalpy changes of reactions are normally given under standard conditions—that is, as ΔH values.

You must remember and understand the definitions of the various types of enthalpy changes of reactions in order to be able to solve problems in this area.

The *standard enthalpy of formation of a compound* (ΔH) is the enthalpy change when one mole of the compound is formed from its elements; each substance taking part in the reaction being in its standard state, i.e. at a pressure (or partial pressure) of 1.013×10^5 Pa. For example, the standard enthalpies of formation of $H_2O(g)$ and $NaHSO_4(s)$ are respectively the enthalpy changes at 1.013×10^5 Pa (i.e. 'atmospheric pressure') corresponding to the following reactions:

$$H_2(g) + \tfrac{1}{2}O_2(g) \rightarrow H_2O(g)$$

$$Na(s) + \tfrac{1}{2}H_2(g) + S(s) + 2O_2(g) \rightarrow NaHSO_4(s)$$

The standard enthalpy changes for other types of reaction are defined in a similar way. For example, the standard enthalpy of combustion of a substance is the enthalpy change when one mole of the substance is burnt completely in oxygen; each substance taking part in the reaction being in its standard state.

Table 9.1 Conventional standard states at any temperature

Standard state for	Definition
(a) a pure gas	gas at 1.013×10^5 Pa
(b) a gas in a mixture	each gas at partial pressure of 1.013×10^5 Pa
(c) a pure liquid or pure solid	liquid or solid when external pressure is 1.013×10^5 Pa
(d) a solute in a solution	solute at unit concentration when external pressure is 1.013×10^5 Pa

Table 9.2 Standard enthalpies of formation
($\Delta H^{\ominus}$) of some compounds at 298 K

Compound	$\Delta H^{\ominus}/\text{kJ mol}^{-1}$
$H_2O(g)$	-241.6
$H_2O(l)$	-285.6
$SO_2(g)$	-296.8
$H_2SO_4(l)$	-811.0
$H_2SO_4(aq)$	-867.6
$NO(g)$	$+90.31$
$NO_2(g)$	$+33.86$
$N_2O_4(g)$	$+13.29$
$NH_3(g)$	-45.98
$CO(g)$	-110.4
$CO_2(g)$	-393.1
$CH_4(g)$	-74.78
$C_2H_6(g)$	-84.45
$C_2H_5OH(l)$	-277.4
$C_3H_8(g)$	-112.7
$Al_2O_3(s)$	-1664
$Fe_2O_3(s)$	-836.0
$C_8H_{18}(l)$	-208.3
$N_2H_4(l)$	$+50.63$

Example 9.3 To illustrate the principle of additivity of enthalpies of reaction

The standard enthalpies of combustion of carbon to carbon monoxide, and of carbon monoxide to carbon dioxide are -110.4 kJ mol^{-1} and -282.7 kJ mol^{-1} respectively. Calculate the standard enthalpy of combustion of carbon to carbon dioxide.

Solution

Step 1 Clarify and define the problem

We want the standard enthalpy of combustion ($\Delta H^{\ominus}$) of carbon to carbon dioxide which is the enthalpy change when carbon is burnt to carbon dioxide under standard conditions.

Step 2 Select the key equation

As key equation we write the balanced equation for the reaction for which $\Delta H^{\ominus}$ is required.

$$C(s) + O_2(g) \rightarrow CO_2(g) \qquad (1)$$

Step 3 Derive the equation for the calculation

We want to calculate $\Delta H^{\ominus}$ for reaction (1). This can be done from the given data and by making use of the principle of additivity of enthalpy changes of reaction, which is an extension of the law of conservation of energy. The given data could be represented in equation form as follows:

$$C(s) + \tfrac{1}{2}O_2(g) \rightarrow CO(g); \quad \Delta H_2^\ominus = -110.4 \text{ kJ mol}^{-1} \tag{2}$$

$$CO(g) + \tfrac{1}{2}O_2(g) \rightarrow CO_2(g); \quad \Delta H_3^\ominus = -282.7 \text{ kJ mol}^{-1} \tag{3}$$

It is seen that equation (1) is the sum of equations (2) and (3); it follows therefore (see note below) that the enthalpy change for reaction (1) would be equal to the sum of the enthalpy changes of reaction (2) and (3). That is

$$\Delta H^\ominus = \Delta H_2^\ominus + \Delta H_3^\ominus \tag{4}$$

Step 4 Collect the data, check the units, and calculate

$$\Delta H^\ominus = \Delta H_2^\ominus + \Delta H_3^\ominus$$

$$= -110.4 \text{ kJ mol}^{-1} + (-282.7 \text{ kJ mol}^{-1})$$

$$= -393.1 \text{ kJ mol}^{-1}$$

The standard enthalpy of combustion of carbon to carbon dioxide is therefore $-393.1 \text{ kJ mol}^{-1}$.

Step 5 Review and check the solution

Note

Recognize that the principle used in the calculation is simply the principle of additivity of enthalpy changes of reactions. We calculate $\Delta H^\ominus$ for reaction (1) by adding together the $\Delta H^\ominus$ values for reactions (2) and (3).

Recognize also that to do the calculation it is necessary to represent all the reactions involved in equation form (see equations 1, 2, and 3); only then can we see how the known enthalpies of reactions have to be made use of to obtain the enthalpy change for the required reaction.

Example 9.4 Prediction of enthalpies of reactions

The standard enthalpy of combustion of octane (C_8H_{18}; octane is the main constituent of petrol) is $-5.45 \times 10^3 \text{ kJ mol}^{-1}$. The standard enthalpies of formation of CO_2 and H_2O are $-393.1 \text{ kJ mol}^{-1}$ and $-285.6 \text{ kJ mol}^{-1}$ respectively. Predict the standard enthalpy of formation of octane.

Solution

Step 1 Clarify and define the problem

We want the standard enthalpy of formation of octane. Let us represent this by $\Delta H^\ominus$. A diagram may help to clarify the problem (see the cycle at the end of this solution).

Step 2 Select the key equation

The required quantity—the standard enthalpy of formation of octane—is the $\Delta H^\ominus$ value for the reaction

$$8C(s) + 9H_2(g) \rightarrow C_8H_{18}(1) \tag{1}$$

Step 3 Derive the equation for the calculation

We can find $\Delta H^{\ominus}$ for the above reaction by making use of the principle of additivity of enthalpy changes of reactions. The given data can be represented in equation form as follows:

$$C_8H_{18}(1) + 12.5O_2(g) \rightarrow 8CO_2(g) + 9H_2O(1), \quad \Delta H_2^{\ominus} = -5450 \text{ kJ mol}^{-1} \quad (2)$$

$$C(s) + O_2(g) \rightarrow CO_2(g), \quad \Delta H_3^{\ominus} = -393.1 \text{ kJ mol}^{-1} \quad (3)$$

$$H_2(g) + 0.5O_2(g) \rightarrow H_2O(1), \quad \Delta H_4^{\ominus} = -285.6 \text{ kJ mol}^{-1} \quad (4)$$

The problem essentially is—how can we make use of equations (2), (3), and (4) so as to obtain $\Delta H^{\ominus}$ for reaction (1)?

Let us first compare equations (2) and (1); C_8H_{18} appears in both equations. In equation (1) C_8H_{18} appears on the right side; so we rewrite equation (2) to make C_8H_{18} appear on the right side. Reversing equation (2) means that we must reverse the sign of $\Delta H_2^{\ominus}$. We then obtain

$$8CO_2(g) + 9H_2O(g) \rightarrow C_8H_{18}(1) + 12.5O_2(g), \quad \Delta H_5^{\ominus} = +5450 \text{ kJ mol}^{-1} \quad (5)$$

Now look at equations (3) and (1); C appears in both but in different amounts. To make C in equation (3) tally with that in equation (1) we multiply (3) by 8; we then obtain

$$8C(s) + 8O_2(g) \rightarrow 8CO_2(g), \quad \Delta H_6^{\ominus} = -8 \times 393.1 \text{ kJ mol}^{-1} \quad (6)$$

Similarly to make H_2 in equation (4) tally with that in equation (1) we have to multiply equation (4) by 9.

$$9H_2(g) + 4.5O_2(g) \rightarrow 9H_2O(1), \quad \Delta H_7^{\ominus} = -9 \times 285.6 \text{ kJ mol}^{-1} \quad (7)$$

It is now seen that on adding equations (5), (6), and (7) we obtain equation (1). On adding the $\Delta H^{\ominus}$ values for reactions (5), (6), and (7) we would therefore obtain $\Delta H^{\ominus}$ for reaction (1). That is

$$\Delta H^{\ominus} = \Delta H_5^{\ominus} + \Delta H_6^{\ominus} + \Delta H_7^{\ominus} \quad (8)$$

Step 4 Collect the data, check the units, and calculate

$$\Delta H_5^{\ominus} = +5450 \text{ kJ mol}^{-1} \text{ (see equation 5)}$$

$$\Delta H_6^{\ominus} = -3144.8 \text{ kJ mol}^{-1} \text{ (see equation 6)}$$

$$\Delta H_7^{\ominus} = -2570.4 \text{ kJ mol}^{-1} \text{ (see equation 7)}$$

By equation (8) we therefore have

$$\Delta H^{\ominus} = (5450 \text{ kJ mol}^{-1}) + (-3144.8 \text{ kJ mol}^{-1}) + (-2570.4 \text{ kJ mol}^{-1})$$

$$= -265.2 \text{ kJ mol}^{-1}$$

The standard enthalpy of formation of octane is therefore $-265.2 \text{ kJ mol}^{-1}$.

Step 5 Review and check the solution

Note

The enthalpy of formation of octane cannot be determined directly by experiment because carbon and hydrogen do not react to give octane. It can, however, be calculated as illustrated in the above example. This calculation is based on the principle (also known as Hess's Law) that ΔH values for reactions can be obtained by the addition of ΔH values for appropriate reactions. Such calculations are particularly useful for reactions where the ΔH required cannot be measured experimentally.

To clarify a Hess's law problem it is sometimes helpful to draw a diagram showing all the appropriate reactions properly balanced. For this example:

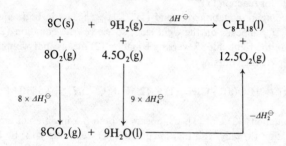

This is known as a thermochemical cycle and it shows two hypothetical routes to the same compound, in this example octane. The enthalpy changes for these two routes can then be equated

$$\Delta H^{\ominus} = 8 \times \Delta H_3^{\ominus} + 9 \times \Delta H_4^{\ominus} + (-\Delta H_2^{\ominus})$$

Example 9.5   Calculation of the enthalpy of dissociation of a weak acid

The enthalpy of neutralization of a very weak monobasic acid HA by NaOH is -54.2 kJ mol^{-1}. For the neutralization of a strong monobasic acid by NaOH ΔH is -57.0 kJ mol^{-1}. Calculate the enthalpy of dissociation of the weak acid HA.

Solution

Step 1   Clarify and define the problem

Let the required quantity—the enthalpy of dissociation of the weak acid HA—be denoted by ΔH.

Step 2   Select the key equation

As key equation we select the balanced equation for the dissociation reaction which is:

$$HA(aq) \rightarrow H^+(aq) + A^-(aq); \quad \Delta H \tag{1}$$

Step 3   Derive the equation for the calculation

We want ΔH for the dissociation reaction (1). This could be obtained from the data given by making use of the principle of additivity of enthalpies of reactions.

The data given for the neutralizations of the weak acid HA and of a strong monobasic acid by OH^- ions may be represented by the equations (see note on page 46)

$$HA(aq) + OH^-(aq) \rightarrow H_2O(l) + A^-(aq); \quad \Delta H_2 = -54.2 \text{ kJ mol}^{-1} \quad (2)$$

$$H^+(aq) + OH^-(aq) \rightarrow H_2O(l); \quad \Delta H_3 = -57.0 \text{ kJ mol}^{-1} \quad (3)$$

How can we relate equation (1) to equations (2) and (3)? In equation (3) H^+ appears on the left side; to make it appear on the right side, we write the reverse reaction of equation (3), remembering to change the sign of ΔH_3.

$$H_2O(l) \rightarrow H^+(aq) + OH^-(aq); \quad \Delta H_4 = +57.0 \text{ kJ mol}^{-1} \quad (4)$$

It is now seen that addition of equations (2) and (4) would give equation (1). ΔH for reaction (1) is therefore, by the principle of additivity, equal to the sum of the ΔH values for reactions (2) and (4). That is

$$\Delta H = \Delta H_2 + \Delta H_4 \quad (5)$$

Step 4 Collect the data, check the units, and calculate

$$\Delta H = \Delta H_2 + \Delta H_4$$
$$= -54.2 \text{ kJ mol}^{-1} + 57.0 \text{ kJ mol}^{-1}$$
$$= +2.8 \text{ kJ mol}^{-1}$$

The heat of dissociation of the weak acid HA is therefore $+2.8 \text{ kJ mol}^{-1}$.

Step 5 Review and check the solution

Note

All strong acids (e.g. HCl, HNO_3) and all strong bases (e.g. NaOH, KOH) are fully dissociated in aqueous solution. The neutralization reaction of a strong acid (say HA) with a strong base (say MOH) could therefore be written as

$$H^+(aq) + A^-(aq) + M^+(aq) + OH^-(aq) \rightarrow H_2O(l) + A^-(aq) + M^+(aq)$$

The ions M^+ and A^- are present on both sides of the equation; they do not change during the reaction and can therefore be cancelled. They are called 'spectator' ions. The only reaction taking place during the neutralization of any strong acid with any strong base is therefore

$$H^+(aq) + OH^-(aq) \rightarrow H_2O(l)$$

The enthalpy of neutralization of any strong acid with any strong base would therefore have the same value, it is equal to the enthalpy of reaction between H^+ ions and OH^- ions to form H_2O its value is $-57.0 \text{ kJ mol}^{-1}$.

For solving the exercises given below, make use of the data given in *Table 9.2* whenever necessary.

Exercise 9.5

Calculate the standard enthalpy of combustion at 298 K of methane.

Exercise 9.6

An important chemical reaction taking place in plants is the chlorophyll catalysed *photosynthesis* of glucose according to the equation

$$6CO_2(g) + 6H_2O(l) \rightarrow C_6H_{12}O_6(s) + 6O_2(g); \quad \Delta H = +403 \text{ kJ mol}^{-1}$$

(a) Is this reaction exothermic or endothermic?
(b) If 1.00 g of CO_2 is converted into glucose, what is the enthalpy change?
(c) If 100 grams of glucose are eaten, how much energy can this provide the body on *respiration*?

(*Note:* the chemical reaction occurring during respiration is the reverse of that occurring during photosynthesis.)

Exercise 9.7

Data on enthalpies of reactions (ΔH) are of importance not only theoretically (for understanding the process) but also practically. One important use of data on enthalpies of reactions is in the evaluation of the quality of fuels. A good fuel is one which releases a large amount of energy when it is burnt.

Make any necessary calculations and arrange the following fuels in an order, starting with the 'best' fuel (compare for the same *mass* of fuel) (a) carbon (b) hydrogen (c) carbon monoxide CO (d) octane (C_8H_{18}; main component of petrol).

Exercise 9.8

Look at the data in Table 9.2 and explain why hydrazine (N_2H_4) is used as a rocket fuel.

Exercise 9.9

One method for the manufacture of iron is by the 'thermite process', which is the reaction between iron(III) oxide and metallic aluminium. The reaction taking place, $Fe_2O_3 + 2Al \rightarrow Al_2O_3 + 2Fe$, is highly exothermic and produces hot molten iron in a few seconds. Calculate ΔH for the reaction.

Exercise 9.10

Estimate the amount of electrical energy necessary to decompose 1.00 g of water into hydrogen and oxygen.

For how long must a current of 1.00 A be passed to decompose electrolytically 1.00 g of water? The voltage applied is 1.80 V (*Note:* joule = volt × ampere × second).

Exercise 9.11

On average a person requires about 12 000 kJ of energy daily for leading a normal life. How many kilograms of sugar (sucrose, $C_{12}H_{22}O_{11}$) must be consumed if all the energy requirements are to be obtained from sugar? The enthalpy of combustion of sugar is -5670 kJ mol^{-1}.

Exercise 9.12

The enthalpy of *solution* of $CaCl_2 \cdot 6H_2O$ is $+14.7$ kJ mol^{-1}. The enthalpy of *hydration* of $CaCl_2 \cdot 6H_2O$ is -97.4 kJ mol^{-1}. Calculate the enthalpy of *solution* of $CaCl_2$. (*Note:* the enthalpy of solution of a substance is the heat change when the substance is dissolved in excess of solvent. The enthalpy of hydration of a substance (e.g. $CaCl_2 \cdot 6H_2O$) is the enthalpy change when that substance is formed from the corresponding anhydrous substance (e.g. $CaCl_2$).)

Exercise 9.13

The standard enthalpies of formation of $CuSO_4 \cdot 5H_2O(s)$ and of $H_2O(l)$ are given. What further data would you require to calculate the standard enthalpy of hydration of $CuSO_4 \cdot 5H_2O(s)$?

Exercise 9.14

Consider the data given in Example 9.5. Suppose that the dissociation of the acid HA, instead of being negligibly small (as was implicitly assumed in the solution given in Example 9.5), is 50% in the weak acid solution used. Show that the enthalpy of dissociation of the weak acid would then be $+5.6$ kJ mol^{-1}.

Review of Chapter 9

In this chapter we considered the first law of thermodynamics and enthalpies of reactions (thermochemistry). The first law, which is an extension of the law of conservation of energy, states that for any process the internal energy change (ΔU) depends *only* on the initial and final states of a system. In equation form the first law is given as

$$\Delta U = q + w \qquad\qquad \textbf{9.1}$$

where q and w are the heat and mechanical energy supplied to the system.

Most of the calculations in thermochemistry are based on the application of the principle of additivity of enthalpy changes of reactions. This principle enables us to add known ΔH values of certain reactions so as to predict ΔH values for other reactions.

10 E.m.f. and electrolysis

10.1 Electrical quantities and units

To understand Chapters 10 and 11 some background knowledge about certain electrical quantities is required; this is explained in this section.

There is only one *basic* electrical quantity (see section 2.4); this is current (I). The SI unit for current is ampere (A).

Four common derived electrical quantities are charge (Q), potential (V), resistance (R), and conductance (G).

The defining equations for these quantities together with their SI units, and the symbols for SI units, are given in Table 10.1. The information given in this table should be memorized.

Table 10.1 Defining equations and units for some electrical quantities

Electrical quantity	Defining equation	SI unit	Symbol for SI unit
current (I)		ampere	A
charge (Q)	$Q = I \times t$	coulomb	C
potential (V)	$V = \dfrac{\text{Energy}}{Q}$	volt	V
resistance (R)	$R = \dfrac{V}{I}$	ohm	Ω
conductance (G)	$G = \dfrac{1}{R}$	siemens	S

Exercise 10.1

Make use of the information given in Table 10.1 to calculate, in SI units:

(a) the charge that flows through a circuit when a current of 0.200 A flows for 3.00 hours
(b) the change in electrical energy when a charge of 9.649×10^4 C, which is the charge present on one mole of electrons, flows through an electrical conductor from a potential of 1.000 V to 0.500 V
(c) the conductance of a column of solution if 0.100 A flows through it when the potential difference applied is 0.500 V.

Exercise 10.2

Derive the relationship between the basic SI units and the SI unit for (a) charge (b) potential (c) resistance (d) conductance.

10.2 Electrode potentials

At an electrode a reversible reaction takes place. This reaction differs from a chemical reaction in that electrons appear in the equation for the reaction. An electrode reaction is always a half-reaction of an oxidation-reduction reaction (see section 4.4).

Some of the common electrodes, and the reactions taking place at them, are tabulated in Table 10.2. Study this table carefully and note that the reversible reaction taking place is always an *oxidation-reduction reaction*. The forward reactions in Table 10.2 are reduction reactions while the reverse reactions are oxidations. You will recall from Chapter 4 that an oxidation reaction is one in which electrons are released while a reduction is one in which electrons are accepted.

Consider the reduction reaction taking place at an electrode; let us write this, in general form, as

$$aA + bB + ze \rightarrow xX + yY$$

The electrode potential E, which is defined as the potential corresponding to the *reduction* reaction, can be shown to be related to the concentrations of the species involved in the electrode reaction by the equation, known as the Nernst equation

$$E = E^{\ominus} - \frac{2.303RT}{zF} \log_{10} \frac{[X]^x[Y]^y}{[A]^a[B]^b} \qquad \textbf{Equation 10.1}$$

R is the gas constant, T is the temperature, F is the Faraday constant (9.649×10^4 C mol^{-1}), and $E^{\ominus}$ is the electrode potential when each of the species involved in the electrode reaction is in its standard state (for definition of standard states, see Table 9.1); it is known as the standard electrode potential.

Electrode potential values are relative values; remember that they are all based on the convention that for the hydrogen electrode $E^{\ominus} = 0$.

Look at **Equation 10.1**. It shows that the electrode potential under any conditions (E) is different from the electrode potential when each substance is in its standard state ($E^{\ominus}$) by the third term which is $(2.303\ RT/zF) \log_{10} \{[X]^x[Y]^y/[A]^a[B]^b\}$. This term is evidently a correction factor which takes into account the influence on the electrode potential due to substances being not in their standard states. It is, therefore, clear that the third term in **Equation 10.1** should include only the concentrations of those substances that are not present in their standard states. In other words, when we write down **Equation 10.1** for any electrode reaction, we should exclude the concentration term of a substance if the substance is present in its standard state. For example, for the hydrogen electrode, hydrogen gas is in its standard state and so **Equation 10.1** becomes

$$E = E^{\ominus} - \frac{2.303RT}{zF} \log_{10} \frac{1}{[H^+]}$$

Table 10.2   Standard Electrode Potentials ($E^\ominus$) at 298.2 K. $E^\ominus$ is the electrode potential when each substance involved in the reaction is in its standard state (see Table 9.1).

Electrode	Electrode reaction	$E^\ominus$/Volt
$Pt\|S_2O_8^{2-}, SO_4^{2-}$	$S_2O_8^{2-}(aq) + 2e = 2SO_4^{2-}(aq)$	+1.98
$Au^{3+}\|Au$	$Au^{3+}(aq) + 3e = Au(s)$	+1.50
$Pt, Cl_2\|Cl^-$	$Cl_2(g) + 2e = 2Cl^-(aq)$	+1.36
$Pt\|Cr_2O_7^{2-}, Cr^{3+}$	$Cr_2O_7^{2-}(aq) + 14H^+(aq) + 6e = 2Cr^{3+}(aq) + 7H_2O$	+1.33
$Pt, Br_2\|Br^-$	$Br_2(1) + 2e = 2Br^-(aq)$	+1.07
$Ag^+\|Ag$	$Q(aq) + 2H^+(aq) + 2e = H_2Q(aq)$	+0.70
$Hg\|Hg_2Cl_2, Cl^-$†	$Ag^+(aq) + e = Ag(s)$	+0.80
$Pt\|Fe^{3+}, Fe^{2+}$	$Hg_2Cl_2(s) + 2e = 2Hg(1) + 2Cl^-(aq)$	+0.77
$Pt\|Q, H_2Q$*	$Fe^{3+}(aq) + e = Fe^{2+}(aq)$	+0.40
$Pt, O_2\|OH^-$	$O_2(g) + 2H_2O + 4e = 4OH^-(aq)$	+0.34
$Cu^{2+}\|Cu$	$Cu^{2+}(aq) + 2e = Cu(s)$	+0.27
$Ag\|AgCl, Cl^-$	$AgCl(s) + e = Ag(s) + Cl^-(aq)$	+0.22
$H^+\|H_2, Pt$	$2H^+(aq) + 2e = H_2(g)$	0.000
$Zn^{2+}\|Zn$	$Zn^{2+}(aq) + 2e = Zn(s)$	−0.76
$Na^+\|Na$	$Na^+(aq) + e = Na(s)$	−2.71
$Li^+\|Li$	$Li^+(aq) + e = Li(s)$	−3.05

* Q is quinone (i.e. cyclohexadiene—1,4-dione) and QH_2 is quinol (i.e. benzene—1,4-diol). The system is called the quinhydrone electrode.

† This is the calomel electrode.

Exercise 10.3

Show that the electrode potentials for the first three electrode reactions in Table 10.2 would be given respectively by the following equations:

$$E = E^\ominus - \frac{2.303RT}{2F} \log_{10} \frac{[SO_4^{2-}]^2}{[S_2O_8^{2-}]}$$

$$E = E^\ominus + \frac{2.303RT}{3F} \log_{10} [Au^{3+}]$$

$$E = E^\ominus - \frac{2.303RT}{2F} \log_{10} [Cl^-]^2$$

Example 10.1   Calculation of ionic concentration from electrode potential

Calculate the hydrogen ion concentration in a solution in which a hydrogen electrode has an electrode potential −0.0852 V at 298.2 K. (Always assume, unless specifically stated otherwise, that a gas in an electrode system is in its standard state.)

Solution

Step 1   Clarify and define the problem

To calculate the concentration of an ion from electrode potential data we have to think of an equation which relates these two quantities. This equation is always obtained by the application of **Equation 10.1** to the equation for the appropriate electrode reaction.

Step 2 Select the equation

The relationship between the hydrogen ion concentration $[H^+]$ and the electrode potential E of a hydrogen electrode can be obtained by the application of **Equation 10.1**. To apply this equation it is necessary first to write down the reduction reaction occurring at the electrode; at a hydrogen electrode this is

$$H^+(aq) + e \rightarrow \tfrac{1}{2}H_2(g)$$

Application of **Equation 10.1** to the above reaction then shows that (we assume that hydrogen gas is in its standard state; its concentration does not therefore appear in the equation below)

$$E = E^\ominus - \frac{2.303RT}{zF} \log_{10} \frac{1}{[H^+]} \qquad \text{(1) \textbf{Equation 10.1}}$$

Step 3 Derive the equation for the calculation

Equation (1) could be used for the calculation of the required quantity, $[H^+]$. On rearranging equation (1) we get

$$E = E^\ominus + \frac{2.303RT}{zF} \log_{10} [H^+]$$

$$\therefore \qquad \log_{10}[H^+] = \frac{(E - E^\ominus)\, zF}{2.303RT}$$

$$\therefore \qquad [H^+] = \text{antilog} \frac{(E - E^\ominus)\, zF}{2.303RT} \qquad (2)$$

Step 4 Collect the data, check the units, and calculate

We can calculate $[H^+]$ by making use of equation (2).

$E = -0.0852$ V (from data)

$E^\ominus = 0$ (for the hydrogen electrode—by convention)

$T = 298.2$ K (from data)

$z = 1$

$F = 9.649 \times 10^4$ C mol^{-1}

$R = 8.314$ J K^{-1} mol^{-1}

$$\therefore \qquad [H^+] = \text{antilog} \frac{(-0.0852 \text{ V})(9.649 \times 10^4 \text{ C mol}^{-1})}{2.303 \times 8.314 \text{ J K}^{-1} \text{mol}^{-1} \times 298.2 \text{ K}}$$

$$= \text{antilog} \, (-1.444) \qquad \qquad \text{(see note 1)}$$

$$= \text{antilog} \, (\bar{2}.556)$$

$$= 3.60 \times 10^{-2} \text{ mol dm}^{-3} \qquad \qquad \text{(see note 2)}$$

Step 5 Review and check the solution

Notes

1 From section 10.1, it is clear that J/C = V and all units would therefore cancel out.
2 The unit for concentration in the answer has been given as mol dm^{-3}. This conclusion follows from the fact that this unit had already been 'built in' into the value of $E^\ominus$. To understand this statement analyse, for example, the equation $E = E^\ominus + (2.303RT/zF) \log_{10}[H^+]$. It is clear from this equation that $E = E^\ominus$ when $[H^+] = 1$, whatever the units of $[H^+]$. The actual value of $E^\ominus$ would therefore depend on the units in which $[H^+]$ is expressed. The $E^\ominus$ values in the literature (see Table 10.2) are for the case when concentrations are given in mol dm^{-3}. The concentration terms in the equation for E should therefore be in mol dm^{-3}.

Example 10.2 Calculation of electrode potential

25.00 cm^3 of an acidified solution of 0.1000 M Fe^{2+} ions is titrated with 0.0500 M potassium dichromate VI (K$_2$Cr$_2$O$_7$).

(a) After the addition of 2.00 cm^3 of the K$_2$Cr$_2$O$_7$ solution calculate the potential of a platinum wire dipped into this solution.
(b) What volume of the K$_2$Cr$_2$O$_7$ solution must be added to make the electrode potential of the Pt|Fe^{3+}, Fe^{2+} electrode equal to its $E^\ominus$ value?
$E^\ominus$ for Pt|Fe^{3+}, Fe^{2+} is +0.771 V, at T, 298.2 K.

Solution to Part (a)

Step 1 Clarify and define the problem

When K$_2$Cr$_2$O$_7$ is added it oxidizes a part of the Fe^{2+} ions present in the solution to Fe^{3+}. A platinum wire dipped into the solution would therefore constitute a Pt|Fe^{3+}, Fe^{2+} electrode. Our problem, therefore, is to calculate the electrode potential E of a Pt|Fe^{3+}, Fe^{2+} electrode.

Step 2 Select the key equation

We have to calculate E for a Pt|Fe^{3+}, Fe^{2+} electrode; $E^\ominus$ is given in the data. As key equation we therefore select the equation which relates E and $E^\ominus$. This equation is obtained by applying **Equation 10.1** to the *reduction reaction* occurring at the Pt|Fe^{3+}, Fe^{2+} electrode which is

$$Fe^{3+}(aq) + e \rightarrow Fe^{2+}(aq)$$

$$\therefore \qquad E = E^\ominus - \frac{2.303RT}{zF} \log_{10} \frac{[Fe^{2+}]}{[Fe^{3+}]} \qquad \text{(1) Equation 10.1}$$

Step 3   Derive the equation for the calculation

Equation (1) shows that to calculate E we require the concentrations of Fe^{2+} and Fe^{3+} in the solution. The concentrations of these ions, after the addition of 2.00 cm^3 of 0.0500 M $K_2Cr_2O_7$ solution to 25.00 cm^3 of a solution containing 0.1000 M Fe^{2+} ions, could be calculated by making use of

(i) **Equation 4.2,** $c = \dfrac{n}{V}$

(ii) the information provided by the balanced equation for the reaction between Fe^{2+} ions and $K_2Cr_2O_7$. We then obtain (do this calculation as an exercise; if you find any difficulty study section 4.4 once again)

$$[Fe^{2+}] = 0.0704 \text{ mol dm}^{-3}$$

$$[Fe^{3+}] = 0.0222 \text{ mol dm}^{-3}$$

Equation (1) can therefore be rewritten as

$$E = E^{\ominus} - \frac{2.303\,RT}{zF} \log_{10}\left(\frac{0.0704}{0.0222}\right) \qquad (2)$$

Equation (2) can be used for the calculation of E.

Step 4   Collect the data, check the units, and calculate

$$E^{\ominus} = +0.771 \text{ V} \qquad \text{(from data)}$$

$$T = 298.2 \text{ K} \qquad \text{(from data)}$$

$$R = 8.314 \text{ J K}^{-1}\,\text{mol}^{-1}$$

$$F = 9.649 \times 10^4 \text{ C mol}^{-1}$$

$$z = 1$$

$$\therefore \quad E = 0.771 \text{ V} - \frac{2.303(8.314 \text{ J K}^{-1}\,\text{mol}^{-1})(298.2 \text{ K})}{9.649 \times 10^4 \text{ C mol}^{-1}} \times \log_{10}\left(\frac{0.0704}{0.0222}\right)$$

$$= 0.771 \text{ V} - 0.0591 \text{ V} \times \log_{10}\left(\frac{0.0704}{0.0222}\right)$$

$$= 0.742 \text{ V}$$

The required potential of the $Pt/Fe^{3+}, Fe^{2+}$ electrode is therefore 0.742 V.

Solution to Part (b)

Steps 1 and 2

These steps are similar to those given in Part (a) and the key equation too is the same and is

$$E = E^{\ominus} - \frac{2.303RT}{zF} \log_{10} \frac{[Fe^{2+}]}{[Fe^{3+}]} \tag{2}$$

Step 3   Derive the equation for the calculation

We have to make the electrode potential E equal to $E^{\ominus}$. To see how this can be done look at equation (2); it shows that E would be equal to $E^{\ominus}$ only when

$$\frac{2.303RT}{zF} \log_{10} \frac{[Fe^{2+}]}{[Fe^{3+}]} = 0 \tag{3}$$

Since R and F are constants and $T = 298.2$ K and $z = 1$, it follows that the only way by which equation the left side in equation (3) can be made equal to zero is to make

$$\frac{[Fe^{2+}]}{[Fe^{3+}]} = 1 \tag{4}$$

Step 4   Collect the data, check the units, and calculate

The question asked can now be reformulated as follows. What volume of a 0.0500 M $K_2Cr_2O_7$ solution must be added to 25.00 cm^3 of a 0.1000 M Fe^{2+} solution so that in the solution obtained $[Fe^{2+}] = [Fe^{3+}]$?

This question is easily answered by a calculation which makes use of the balanced equation for the reaction and the equation $c = n/V$ [Equation 4.2]. Such a calculation (do this calculation as an exercise) would show that

$$V = 8.33 \text{ cm}^3$$

8.33 cm^3 of 0.0500 M $K_2Cr_2O_7$ solution must therefore be added to 25.00 cm^3 of a 0.1000 M Fe^{2+} solution to make the electrode potential of a $Pt|Fe^{3+}, Fe^{2+}$ electrode equal to the $E^{\ominus}$ value.

Step 5   Review and check the solution

For $E^{\ominus}$ values use Table 10.2. Assume that all data given are at 298.2 K, unless otherwise specified.

Exercise 10.4

Show that the electrode potential of the silver–silver chloride electrode, $Ag(s)|AgCl(s), Cl^-(aq)$, depends only on the chloride ion concentration.

Exercise 10.5

Calculate the standard electrode potential of the silver–silver chloride electrode $Ag(s)|AgCl(s),Cl^-(aq)$, given that its electrode potential in a 0.03000 mol dm^{-3} solution of chloride ions at 25.0 °C is 0.3125 V.

Exercise 10.6

Calculate the silver ion concentration in a solution in which a silver electrode has a potential of +0.666 V at 25.0 °C.

Exercise 10.7

Calculate the chloride ion concentration of a solution in which a chlorine electrode has an electrode potential +1.370 V. What will be the electrode potential of a calomel electrode, $Hg(1)|Hg_2Cl_2(s)|Cl^-(aq)$, dipped in this solution?

Exercise 10.8

Calculate the electrode potential of (a) a hydrogen electrode (b) an oxygen electrode Pt, $O_2|OH^-(aq)$, dipped in a 1.00 M aqueous solution of (i) NaCl (ii) HCl (iii) NaOH.

Exercise 10.9

A common electrode used for measuring the pH of a solution is the *quinhydrone electrode*. Quinhydrone is a *equimolar* mixture of quinone ($C_6H_4O_2$, cyclohexadiene-1,4-dione, abbreviated Q) and hydroquinone ($C_6H_4(OH)_2$, benzene-1,4-diol, abbreviated QH_2). The reversible electrode reaction taking place is

$$Q(aq) + 2H^+(aq) + 2e \rightleftharpoons H_2Q(aq)$$

Calculate the pH of a solution in which a quinhydrone electrode had an electrode potential + 0.650 V.

(Hint: recognize that since quinhydrone is an equimolar mixture of Q and H_2Q, the concentrations, in solution, of Q and H_2Q are equal.)

10.3 Electrochemical cells

When two electrodes are joined together, an electrochemical cell is obtained. For example, if a zinc electrode ($Zn^{2+}(aq)|Zn$) and a hydrogen electrode (Pt, $H_2(g)|H^+(aq)$) are joined together the cell obtained could be written as

$$Pt, H_2(g)|H^+(aq)|Zn^{2+}(aq)|Zn(s)$$

The e.m.f. E of an electrochemical cell is related to the electrode potentials of the two electrodes of the cell by the equation

$$E = E_R - E_L \qquad\qquad \textbf{Equation 10.2}$$

where E_R and E_L are respectively the electrode potentials of the right electrode and left electrode in the cell. (The terms right electrode and left electrode refer to the cell as written; and with reduction at the right electrode and oxidation at the left electrode).

The electrochemical cell reaction is merely the sum of the two electrode reactions, one of which is oxidation while the other is reduction. The relationship between the e.m.f. of the cell and the concentrations of the substances taking part in the cell reaction is also given by **Equation 10.1**.

The sign of the e.m.f. of a cell is for the cell as *written down*. Consider, for example, the cell

$$\text{Pt, H}_2(\text{g})|\text{HClaq}|\text{AgCl, Ag} \qquad (1)$$

The standard e.m.f. ($E^\ominus$) of this cell, calculated by the application of **Equation 10.2** to the data given in Table 10.2, would be seen to have a +ve value (+0.22 V).

Example 10.3 Calculations of e.m.f. of an electrochemical cell

Consider the electrochemical cell

$$\text{Zn}|\text{Zn}^{2+}(\text{aq})|\text{H}^+(\text{aq})|\text{H}_2, \text{Pt}$$

Make use of data from Table 10.2 and calculate at 298.2 K the e.m.f. of the cell when the concentrations of $\text{Zn}^{2+}(\text{aq})$ and $\text{H}^+(\text{aq})$ are 0.100 mol dm^{-3} and 0.0100 mol dm^{-3} respectively.

$$\frac{2.303RT}{F} = 0.0591 \text{ V at 298.2 K}$$

Solution

Step 1 Clarify and define the problem

We have to calculate the e.m.f. E of an electrochemical cell given the concentrations of the ions in the solution. To calculate E from the data given we must therefore derive an equation which relates E to the concentrations of the ions.

Step 2 Select the key equation

The required relationship between the e.m.f. E of a cell and the concentrations of the species taking part in the cell reaction is obtained by applying **Equation 10.1** to the cell reaction. We must therefore first know the cell reaction.

In the cell given

$$\text{Zn(s)}|\text{Zn}^{2+}(\text{aq})|\text{H}^+(\text{aq})|\text{H}_2(\text{g}), \text{Pt}$$

it is seen that (read from left to right) Zn becomes Zn^{2+} ions and H^+ becomes hydrogen gas. That is

left electrode reaction is: $\text{Zn(s)} \rightarrow \text{Zn}^{2+}(\text{aq}) + 2\text{e}$ (oxidation)

right electrode reaction is: $2\text{H}^+(\text{aq}) + 2\text{e} \rightarrow \text{H}_2(\text{g})$ (reduction)

∴   cell reaction is: $\text{Zn(s)} + 2\text{H}^+(\text{aq}) \rightarrow \text{Zn}^{2+}(\text{aq}) + \text{H}_2(\text{g})$

Application of **Equation 10.1** to this cell reaction then shows that

$$E = E^\ominus - \frac{2.303RT}{zF} \log_{10} \frac{[\text{Zn}^{2+}]}{[\text{H}^+]^2} \qquad (1) \text{ Equation 10.1}$$

Step 3 Derive the equation for the calculation

We can calculate E by equation (1) provided we know $E^\ominus$. This is not given in the data but it can be calculated by the application of **Equation 10.2**. In the cell considered the right electrode is the hydrogen electrode and therefore the application of **Equation 10.2** shows that

$$E^{\ominus} = E_{H_2}^{\ominus} - E_{Zn}^{\ominus}$$ (2) **Equation 10.2**

where $E_{H_2}^{\ominus}$ and $E_{Zn}^{\ominus}$ refer respectively the standard electrode potentials of the hydrogen electrode and the zinc electrode.

By replacing $E^{\ominus}$ in equation (1) by equation (2) we obtain

$$E = (E_{H_2}^{\ominus} - E_{Zn}^{\ominus}) - \frac{2.303\,RT}{zF}\,\log_{10}\frac{[Zn^{2+}]}{[H^+]^2}$$ (3)

In the problem statement it is also given that $2.303\,RT/F$ is equal to 0.0591 V at 298.2 K. We may therefore simplify equation (3) and write

$$E = (E_{H_2}^{\ominus} - E_{Zn}^{\ominus}) - \frac{0.0591\,\text{V}}{z}\,\log_{10}\frac{[Zn^{2+}]}{[H^+]^2}$$ (4)

Equation (4) can be used to calculate the required quantity E.

Step 4 Collect the data, check the units, and calculate

$$E_{H_2}^{\ominus} = 0$$ (by convention)

$$E_{Zn}^{\ominus} = -0.76$$ (from Table 10.2)

$$[Zn^{2+}] = 0.100 \text{ mol dm}^{-3}$$

$$[H^+] = 0.0100 \text{ mol dm}^{-3}$$

$$z = 2$$

$$\therefore \qquad E = 0 - (-0.76 \text{ V}) - \frac{0.059 \text{ V}}{2}\,\log_{10}\frac{0.100}{(0.0100)^2}$$

$$= 0.76 \text{ V} - 0.088 \text{ V}$$

$$= +0.67 \text{ V}$$

The required e.m.f. of the cell is therefore $+0.674$ V.

Step 5 Review and check the solution

Exercise 10.10

An electrode $M|M^+$ (0.1 M) is joined with a standard hydrogen electrode to form a cell. Experimentally it is then found that the potential difference between the two electrodes is 0.20 V, the hydrogen electrode being the positive electrode. What is the e.m.f. of the cell Pt, $H_2(g)|H^+$ (1.0 M)$|M^+$ (0.1 M)$|M$?

Exercise 10.11

Use the data given in Table 10.2 to calculate $E^{\ominus}$ for each of the following cells.

(a) Pt, $H_2(g)|HCl(aq)|Cl_2(g)$, Pt
(b) Pt, $Cl_2(g)|HCl(aq)|H_2(g)$, Pt
(c) Pt, $H_2(g)|HCl(aq)|AgCl(s)|Ag(s)$
(d) $Pt|Fe^{2+}(aq)$, $Fe^{3+}(aq)|KCl(aq)|Hg_2Cl_2(s)|Hg(1)$

Write the electrode reactions, the cell reaction and the equation showing the dependence of e.m.f. on concentrations for each of the cells given.

Exercise 10.12

Write the electrode reactions and the cell reaction for the cell Pt, $H_2|HBr$ (0.100 M)|$Br_2(1)$, Pt. Calculate the e.m.f. of this cell at 25 °C.

$$E^{\ominus} \text{ for bromine electrode at 25 °C} = +1.07 \text{ V}$$

Exercise 10.13

A student wants to measure the pH of a solution. He makes use of a cell formed by combining a hydrogen electrode and a calomel electrode $Hg(s)|HgCl_2(s)|Cl^-$ (aq).

The potential difference between the two electrodes is found to be 0.339 V, the hydrogen electrode being the negative electrode. The electrode potential of the calomel electrode used is +0.27 V. What would be the pH of the solution? Assume that temperature is 298.2 K.

Exercise 10.14

Consider the electrochemical cell

$$Pt, H_2(g)|HCl(aq)|AgCl(s)|Ag$$

If the HCl solution is at a concentration 0.100 mol dm^{-3} show that at 298.2 K

(a) the electrode potential of the hydrogen electrode is -0.0591 V
(b) the electrode potential of the silver–silver chloride electrode is $+0.281$ V
(c) the e.m.f. of the cell calculated from the electrode potential values obtained above is $+0.340$ V
(d) the e.m.f. of the cell calculated by the application of **Equation 10.1** to the cell reaction is the same as that in (c).

$$E^{\ominus} \text{ for } Cl^-|AgCl, Ag(aq) = +0.222 \text{ V at 298.2 K}$$

Recognize that the e.m.f. of a cell could be calculated by one of two alternative methods:

(i) by first calculating the electrode potentials of the two electrodes and then applying **Equation 10.2** as in (c) above
(ii) by first writing down the cell reaction and then applying **Equation 10.1** as in (d) above.

10.4   Applications of e.m.f. measurements

Most of the applications of electrode potential and e.m.f. measurements are based on the determination of the concentration of ions in a solution by the application of **Equation 10.1**.

Equation 10.1 shows that the electrode potential is quantitatively related to the concentrations of the ions taking part in the electrode reaction. It therefore follows that *by measuring the potential of an appropriate electrode we can determine the concentration of a desired ion in a solution.*

As an example, suppose that we want to find the hydrogen ion concentration in a solution. To do so we must make use of an electrode whose potential depends on the concentration of this ion. A hydrogen electrode is therefore suitable; its potential (E) in the solution is related to the hydrogen ion concentration by the equation (see Example 10.1)

$$E = \frac{2.303 RT}{F} \log_{10}[\text{H}^+]$$

From E we can therefore calculate $[\text{H}^+]$.

Example 10.4   Solubility product from e.m.f.

The e.m.f. of the cell

$$\text{Ag}|\text{AgCl(s), KCl (0.100 M)}|\text{AgNO}_3 \ (1.00 \times 10^{-3} \text{ M})|\text{Ag}$$

is $+0.3426$ V at 25 °C. Calculate the solubility product of silver chloride at 25.0 °C.

$$\frac{2.303 RT}{F} = 0.0591 \text{ V at } 25.0 \text{ °C}$$

Solution

Step 1 Clarify and define the problem
We have to calculate the solubility product of AgCl from e.m.f. data.

Step 2 Select the key equation
We start with the defining equation for the required quantity—the solubility product (K) of AgCl

$$K = [\text{Ag}^+][\text{Cl}^-] \qquad \text{(1) (see section 8.5)}$$

Step 3 Derive the equation for the calculation
To calculate K by equation (1) we must know the concentrations of Ag^+ and Cl^- in a saturated solution of AgCl.

Look at the left electrode of the cell; it contains a saturated solution of AgCl (in 0.100 M KCl). In this solution

$[\text{Cl}^-] = 0.100 \text{ M}$
          (since amount of Cl^- ions due to solubility of AgCl is negligibly small)

To calculate K (by equation 1) we have therefore to calculate only the silver ion concentration (this is due to the solubility of AgCl) present in the solution in the left electrode. Can this be done from the e.m.f. value given? Yes, because the cell e.m.f. can be related to the required silver ion concentration by the application of **Equation 10.1** to the cell reaction.

Let the required quantity—the concentration of Ag^+ in the solution in the left electrode—be denoted $[Ag^+]$. The data indicate that in the right electrode the silver ion concentration is 1.00×10^{-3} mol dm^{-3}. Both electrodes may be considered to be $Ag/Ag^+(aq)$ electrodes; the electrode reactions may therefore be written as

$$\text{left electrode reaction: } Ag \rightarrow Ag^+ \; (c = [Ag^+]) + e$$

$$\text{right electrode reaction: } Ag^+ \; (c = 1.00 \times 10^{-3} \text{ M}) + e \rightarrow Ag$$

$$\therefore \qquad \text{cell reaction: } Ag^+ \; (c = 1.00 \times 10^{-3} \text{ M}) \rightarrow Ag^+ \; (c = [Ag^+])$$

Application of **Equation 10.1** to the cell reaction then shows that

$$E = E^{\ominus} - \frac{2.303RT}{zF} \log_{10} \frac{[Ag^+]}{1.00 \times 10^{-3}} \qquad \text{(2) Equation 10.1}$$

To calculate $[Ag^+]$ by equation (2) we need $E^{\ominus}$ for the cell considered. This could be obtained by the application of **Equation 10.2** which shows that

$$E^{\ominus} = E_R^{\ominus} - E_L^{\ominus} \qquad\qquad \textbf{Equation 10.2}$$

$$= 0 \text{ (since } E_R^{\ominus} = E_L^{\ominus}, \text{ both electrodes being considered}$$

$$\text{to be silver electrodes)}$$

Equation (2) can therefore be rewritten as

$$E = -\frac{2.303RT}{zF} \log_{10} \frac{[Ag^+]}{1.00 \times 10^{-3}} \qquad (3)$$

Step 4 Collect the data, check the units, and calculate

$[Ag^+]$ can now be calculated by equation (3) and therefore the required solubility product K by equation (1).

$$E = +0.3426 \text{ V}$$

$$\frac{2.303RT}{F} = 0.0591 \text{ V}$$

$$z = 1$$

By equation (3) we then obtain

$$E = -0.0591 \text{ V} \log_{10} \frac{[Ag^+]}{1.00 \times 10^{-3}}$$

$$\therefore \qquad [Ag^+] = 1.60 \times 10^{-9} \text{ mol dm}^{-3}$$

The concentration of Ag^+ ions in the saturated solution of $AgCl$ present in the left electrode is therefore 1.60×10^{-9} mol dm^{-3}.

The required solubility product K can now be calculated.

$$K = [\text{Ag}^+][\text{Cl}^-]$$

$$= (1.60 \times 10^{-9} \text{ mol dm}^{-3})(0.100 \text{ mol dm}^{-3})$$

$$= 1.60 \times 10^{-10} \text{ mol}^2 \text{ dm}^{-6}$$

Step 5 Review and check the solution

Check and convince yourself that the approximation made during the calculation that the concentration of Cl^- ions is due only to that formed from KCl is justified. This example illustrates one method for the determination of solubility product. Recognize that the method essentially involves the determination of the concentration of an ion in a saturated solution. In the example considered, the silver ion concentration in a saturated solution of AgCl was determined by making use of a silver electrode. The second electrode used (in this example $\text{Ag}|\text{Ag}^+$ ($c = 1.00$ mol dm^{-3}) could be different from the one used here.

Example 10.5 Potentiometric titration

In the electrochemical cell

$$\text{Ag}|\text{AgCl(s), KCl (solution)}|\text{AgNO}_3 \ (1.00 \times 10^{-3} \text{ M})|\text{Ag}$$

25.00 cm^3 of 0.1000 M KCl solution are present in the left electrode. The KCl solution is titrated with a 0.1000 M AgNO$_3$ solution (present in a burette). Calculate the e.m.f. of the cell after the addition of 10.00 cm^3, 20.00 cm^3, 24.00 cm^3, 24.90 cm^3, 25.00 cm^3, 25.10 cm^3, 26.00 cm^3, 30.00 cm^3, and 40.00 cm^3 of the AgNO$_3$ solution. The solubility product of AgCl at room temperature is 1.60×10^{-10} mol^2 dm^{-6}.

Plot e.m.f. *vs* volume of AgNO$_3$ added and comment on the importance of this graph.

Solution

Steps 1, 2, and 3

The first three steps of the solution given in Example 10.4 are also applicable here and we then have

$$E = -\frac{2.303 \, RT}{zF} \log_{10} \frac{[\text{Ag}^+]}{1.00 \times 10^{-3}} \qquad \text{(1) } \textbf{Equation 10.1}$$

where $[\text{Ag}^+]$ is the concentration of silver ions in the solution in the left electrode.

The equation shows that to calculate E (the required quantity) we must calculate $[\text{Ag}^+]$; this is related to the data given (solubility product K) by the equation

$$[\text{Ag}^+] = \frac{K}{[\text{Cl}^-]} \qquad \text{(2) } \textbf{Equation 7.1}$$

where $[\text{Cl}^-]$ is the concentration of chloride ions in the left electrode.

On replacing $[Ag^+]$ in equation (1) by equation (2) we obtain

$$E = -\frac{2.303RT}{zF} \log_{10} \frac{K}{[Cl^-] \times 1.00 \times 10^{-3}} \tag{3}$$

Step 4 Collect the data, check the units, and calculate

Equation (3) could be used to calculate the required e.m.f. (E) values of the cell, after each addition of $AgNO_3$, because $[Cl^-]$ is easily calculated by making use of the balanced equation for the titration reaction ($AgNO_3(aq) + KCl(aq) \rightarrow AgCl(s) + KNO_3(aq)$).

The calculated values of $[Cl^-]$ in the left electrode, and also the required cell e.m.f. E, after each addition of $AgNO_3$, are given in Table 10.3. Check the calculations as an exercise.

Table 10.3   Calculations for Example 10.5

cm³ 0.1000 M AgNO₃ added	[Cl⁻]/mol dm⁻³	E/volt
10.00	4.29×10^{-2}	+0.3210
20.00	1.11×10^{-2}	+0.2860
24.00	2.04×10^{-3}	+0.2426
24.90	2.02×10^{-4}	+0.1831
25.00	1.26×10^{-5}	+0.1121
25.10	8.00×10^{-7}	+0.0413
26.00	8.17×10^{-8}	-0.0173
30.00	1.77×10^{-8}	-0.0566
40.00	6.93×10^{-9}	-0.0807

We have also been asked to plot e.m.f. E against the volume of titrant. Such a plot, which is known as a potentiometric titration graph, is shown in Fig. 10.1. Observe the shape of the graph. Note that the e.m.f. starts to change very rapidly near the equivalence point of the titration. It is seen that the most rapid change is at the equivalence point, after 25.00 cm³ silver nitrate solution has been added.

The importance of a potentiometric titration graph is that the equivalence point of a titration, and therefore the concentration of a solution, could be determined. To detect the equivalence point one merely reads off from the graph the volume at which the e.m.f. of the cell changes most rapidly.

Step 5 Review and check the solution

Notes

1 Potentiometric titrations provide a useful method for determining ionic concentrations. The principle of the method is the same as that in a direct electrode potential measurement to obtain ionic concentrations. In the direct method, one measures the electrode potential E of an electrode and then uses **Equation 10.1** to calculate the ionic concentration (see Example 10.1). We then require an *accurate*

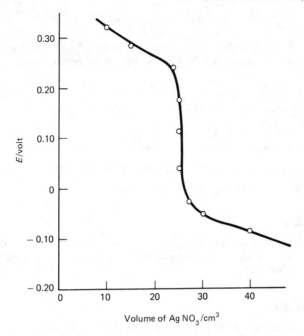

Fig. 10.1   Potentiometric titration graph

value for E. The advantage in a potentiometric titration is that we do not need an accurate value for E. Only the change in E is important. A constant error in the measurement of E would not therefore matter.

2 Potentiometric titrations are versatile and have wide applicability. They are used for acid–base, oxidation–reduction, and precipitation reactions. For example, recognize that the electrochemical cell

$$Pt, H_2(g)|acid\ solution|Cl^-(aq), Hg_2Cl_2(s)|Hg$$

could be used for the potentiometric titration of the acid solution with a standard solution of an alkali.

Exercise 10.15

Look at Table 10.2. List all the electrodes given there that could be used to measure the concentration of chloride ions in a solution.

Exercise 10.16

What electrode would you use, in each case, to determine the concentration of (a) Cu^{2+} ions (b) Br^- ions (c) Fe^{3+} ions, in a solution.

Exercise 10.17

The determination of physical properties such as ionic product, dissociation constant, solubility product, etc. depend basically on the measurements of the relevant ionic concentrations. Indicate how

(a) the ionic product of water
(b) the dissociation constant of CH_3COOH
(c) the solubility product of AgBr
(d) the solubility product of lead iodide
(e) the equilibrium constant for the dissociation

$$Ag(CN)_2^-(aq) \rightleftharpoons Ag^+(aq) + 2CN^-(aq)$$

could be determined by electrode potential measurements.

Exercise 10.18

The e.m.f. of the cell Pt, $H_2(g)$|weak acid HA (0.100 M), KCl|Hg_2Cl_2(s)|Hg is +0.439 V. The electrode potential of the calomel electrode used is +0.27 V. Calculate the dissociation constant of the weak acid.

Exercise 10.19

The e.m.f. of the cell (Ag|AgI, KI (1.00 M)|$AgNO_3$ (0.001 M)|Ag is + 0.720 V. Calculate

(a) the solubility of AgI in 1.00 M KI
(b) the solubility product of AgI.

10.5 Electrolysis

When a solution is electrolysed, electrochemical reactions take place at the two electrodes. At one electrode there is oxidation while at the other there is reduction.

Suppose that a solution containing several ions is electrolysed. More than one reaction may then be possible at each electrode. The reaction that actually takes place could often (not always) be predicted using electrode potential data. The electrode potential, as already explained, is the potential of the *reduction* reaction at an electrode; it is a measure of the ease of the reduction reactio:. The more positive the value of an electrode potential the easier, therefore, would be the reduction reaction. For example, the reduction reaction

$$Ag^+ (c = 1.0 \text{ M}) + e \rightarrow Ag \text{ (for which } E = E^\ominus = +0.80 \text{ V; See Table 10.2)}$$

could be expected to take place more readily than the reduction reaction

$$Cu^{2+} (c = 1.0 \text{ M}) + 2e \rightarrow Cu \text{ (for which } E = E^\ominus = +0.34 \text{ V)}.$$

For oxidation reactions the opposite would be true; the lower or more negative the value of the electrode potential the easier would be the reaction.

Predictions based on electrode potential (E) data as to which ion would discharge are not always valid. This is because an E value indicates only the *minimum* voltage that is theoretically necessary for an electrode reaction; the actual voltage required may be much higher than this minimum because other factors may also affect the rate of an electrode reaction. It is found experimentally that predictions based on E

values are generally true for the discharge of metal ions. For the discharge of other ions, and particularly when a gas is evolved, predictions based on E values are generally found to be unreliable.

Example 10.6 Prediction of cathode reaction

An aqueous solution at 25 °C contains 0.01 M Ag^+ ions and 1.0 M Cu^{2+} ions. What reaction would you expect *first* at the cathode if the applied potential is gradually increased? Use any necessary data from Table 10.2.

Solution

Step 1 Clarify and define the problem

To decide which ion would discharge first, it is necessary to calculate the potentials necessary to discharge each of the ions present, and then compare them. The potential necessary to reduce an ion at an electrode cannot be less than the electrode potential of the corresponding electrode.

Three cathode reactions are possible; these are (remember that H^+ and OH^- ions are always present in any aqueous solution)

$$Ag^+ (0.01\ M) + e \rightarrow Ag(s) \tag{1}$$

$$Cu^{2+} (1.0\ M) + 2e \rightarrow Cu(s) \tag{2}$$

$$H^+ (10^{-7}\ M) + e \rightarrow \tfrac{1}{2}H_2(g) \tag{3}$$

Step 2 Select the key equation

The electrode potentials corresponding to reactions (1), (2), and (3) could be calculated by the application of **Equation 10.1** to electrode reactions (1), (2), and (3) respectively. We then get

$$E_{Ag^+/Ag} = E^{\ominus}_{Ag^+/Ag} - \frac{2.303RT}{F} \log_{10} \frac{1}{[Ag^+]} \tag{4}$$

$$E_{Cu^{2+}/Cu} = E^{\ominus}_{Cu^{2+}/Cu} - \frac{2.303RT}{2F} \log_{10} \frac{1}{[Cu^{2+}]} \tag{5}$$

$$E_{H^+/H_2} = E^{\ominus}_{H^+/H_2} - \frac{2.303RT}{F} \log_{10} \frac{1}{[H^+]} \tag{6}$$

Step 3 Derive the equation for the calculation

Equations (4), (5), and (6) can be used respectively to calculate $E_{Ag^+/Ag}$, $E_{Cu^{2+}/Cu}$ and E_{H^+/H_2} since the required $E^{\ominus}$ values are given in Table 10.2.

Step 4 Collect the data, check the units, and calculate

$$E^{\ominus}_{Ag^+/Ag} = 0.80\ V, \quad [Ag^+] = 0.01\ mol\ dm^{-3}$$

$$E^{\ominus}_{Cu^{2+}/Cu} = 0.34\ V, \quad [Cu^{2+}] = 1.0\ mol\ dm^{-3}$$

$$E^{\ominus}_{H^+/H_2} = 0, \quad\quad\quad [H^+] = 10^{-7}\ mol\ dm^{-3}$$

By the application of equations (4), (5), and (6) we then obtain

$$E_{Ag^+/Ag} = 0.80 - 0.0591 \log_{10} \left(\frac{1}{0.01} \right)$$

$$= 0.80 - 0.12$$

$$= 0.68 \text{ V}$$

$$E_{Cu^{2+}/Cu} = 0.34 - \frac{0.0591}{2} \log_{10} \left(\frac{1}{1.0} \right)$$

$$= 0.34 \text{ V}$$

$$E_{H^+/H_2} = 0 - 0.0591 \log_{10} \left(\frac{1}{10^{-7}} \right)$$

$$= -0.41 \text{ V}$$

The potential for the reduction of Ag^+ (which is the meaning of the term electrode potential) is seen to have the highest value. Ag^+ ions would therefore be expected to discharge first at the negative electrode in the electrolytic cell considered.

Step 5 Review and check the solution

Exercise 10.20

Suggest the likely cathode reaction when an aqueous solution containing 0.10 M Ag^+ ions and 0.10 M Cu^{2+} ions is electrolysed.

Exercise 10.21

A 1.0 M aqueous solution of sodium sulphate is electrolysed. What anode reaction would you expect?

Exercise 10.22

A 1.0 M aqueous solution of copper sulphate is electrolysed using copper electrodes. What anode reaction would you expect?

10.6 Balanced equations for electrode reactions

To calculate the amount of a reactant consumed or of a product formed at an electrode during electrolysis, it is necessary to make use of the balanced equation for the electrode reaction. These calculations are similar to those based on the balanced equation for a chemical reaction (see Chapter 4). However, in an electrode reaction electrons are involved either as a reactant or as a product and, because electrons are charged particles, an electrode reaction always involves a flow of charge. The amount of electrons involved in an electrode reaction is generally indicated not in terms of moles but in terms of charge. To convert charge into moles of electrons, and *vice versa*, is easy since the charge on 1 mole of electrons is 9.649×10^4 C. The charge on one mole of electrons is known as the Faraday constant (F). That is

$$F = 9.649 \times 10^4 \text{ C mol}^{-1}$$

Example 10.7   Amounts involved in an electrode reaction

Calculate the mols of oxygen molecules evolved when a current of 0.200 A is passed through a dilute aqueous solution of sodium hydroxide for 3.00 hours.

Solution

Step 1   Clarify and define the problem

Oxygen is evolved by the oxidation (at the anode) of the OH^- ions present in the aqueous solution. We have to calculate the mols n of O_2 evolved when a known current (I) is passed for a known time (t).

Step 2   Select the key equation

This calculation is based on the information provided by the balanced equation for the reaction. As key equation we therefore select the balanced equation for the electrode reaction which leads to the evolution of oxygen. Application of the laws of conservation of mass and of charge show (see Chapter 4) that the balanced equation is

$$4OH^-(aq) \rightarrow 2H_2O(l) + O_2(g) + 4e \qquad (1)$$

Step 3   Derive the equation for the calculation

Equation (1) shows that when 4 mols of electrons (i.e. a charge of $4 \times 9.649 \times 10^4$ C) flows, 1 mol of O_2 would be evolved.

Therefore when a charge $I \times t$ flows (remember $I \times t = Q$—see Table 10.1) the mols of oxygen (n) that would be evolved is given by

$$n = \frac{1 \text{ mol}}{4 \times 9.649 \times 10^4 \text{ C}} \times (I \times t) \qquad (2)$$

Step 4   Collect the data, check the units, and calculate

$$I = 0.200 \text{ A}$$

$$t = 3.00 \text{ hours}$$

$$= 3.00 \times 60 \times 60 \text{ s}$$

$$\therefore \quad n = \frac{1 \text{ mol} \times 0.200 \text{ A} \times (3.00 \times 60 \times 60 \text{ s})}{4 \times 9.649 \times 10^4 \text{ C}}$$

$$= 5.59 \times 10^{-3} \text{ mol}$$

5.59×10^{-3} mols of oxygen molecules would therefore be evolved when 0.200 A is passed for 3.00 hours through an aqueous solution of sodium hydroxide.

Step 5   Review and check the solution

Exercise 10.23

A widely used method for measuring the total amount of charge that flows through any apparatus (or any circuit) involves connecting a *coulometer* in series with that apparatus. In such an experiment if 1.000 g of silver is deposited at the cathode of a silver coulometer, calculate the charge that flowed through the circuit.

Exercise 10.24

An electric current is passed through two electrolytic cells connected in series, one containing a solution of silver nitrate and the other a solution of sulphuric acid. What volume of oxygen, measured at 25 °C and 750 mmHg, would be liberated from sulphuric acid if (a) 1.000 mol (b) 8.0×10^{22} ions, of Ag^+ are deposited from the silver nitrate solution (1 mm Hg = 1.33×10^2 Pa).

Exercise 10.25

How many coulombs must be passed through 50.00 cm^3 of a concentrated solution of K_2SO_4 so that the concentration of the $S_2O_8^{2-}$ ions formed in the solution is 0.1000 mol dm^{-3}. Assume that the only oxidation reaction taking place is $2SO_4^{2-} \rightarrow S_2O_8^{2-} + 2e$.

Review of Chapter 10

The most important equation in this chapter is

$$E = E^{\ominus} - \frac{2.303RT}{zF} \log_{10} \frac{[X]^x[Y]^y}{[A]^a[B]^b} \qquad \textbf{10.1}$$

For a reaction:

$$aA + bB + ze \rightarrow xX + yY$$

You should understand this equation thoroughly and be able to recall and apply it whenever necessary. This equation is applicable for both electrode reactions and cell reactions. For an electrode reaction E is the electrode potential; for a cell reaction it is the e.m.f.

Most of the applications (and therefore problems) involving e.m.f. and electrode potentials are based on the above equation. Thus if E is measured and the constant $E^{\ominus}$ is known we can calculate concentrations. This is an important general method for determining various ionic concentrations. From concentrations various physical properties (ionic product, dissociation constant, solubility product, etc.) too can be obtained.

The electrode potentials of the two electrodes in a cell are related to the cell e.m.f. E by the equation

$$E = E_R - E_L \qquad \textbf{10.2}$$

where E_R and E_L are the electrode potentials of the right electrode (reduction occurring) and left electrode (oxidation occurring) respectively.

During electrolysis the products formed can often (but not always) be predicted with the help of electrode potential data.

The amounts of reactants consumed or products formed during electrolysis can be calculated by writing the balanced equation for the electrode reaction and making use of the fact that the charge on one mole of electrons is 9.649×10^4 coulombs.

11 Conductance and migration of ions

11.1 Conductivity and molar conductivity

Conductance (G) is defined as the reciprocal of resistance (R). That is

$$G = \frac{1}{R}$$

Equation 11.1

The SI unit for conductance is the siemens, symbol S (see Table 10.1). The conductance of an electrolyte solution is found experimentally to be directly proportional to the area of cross section a of the electrodes and inversely proportional to the length l between the electrodes. That is

$$G = \kappa \frac{a}{l}$$

Equation 11.2

where κ, the constant of proportionality, is known as the *conductivity* of the solution.

The conductivity of an electrolyte is commonly given as *molar conductivity* $\varLambda$; this is defined by the equation

$$\varLambda = \frac{\kappa}{c}$$

Equation 11.3

where c is the concentration of the electrolyte in the solution.

The molar conductivity of an electrolyte is found to increase as the concentration of the electrolyte in the solution decreases. When the concentration tends to zero the molar conductivity is known as the molar conductivity at infinite dilution and is denoted by $\varLambda^\infty$. The SI units for molar conductivity are S m^2 mol^{-1}, but values are more commonly given in units of S cm^2 mol^{-1}.

Example 11.1 Determination of the molar conductivity of an electrolyte in a solution

The resistance of a conductivity cell filled with a 0.1000 M KCl solution is 18.98 Ω; when filled with a 0.1000 M solution of a weak acid HA it is 43.91 Ω. Calculate the molar conductivity of the weak acid in the solution. The conductivity of a 0.1000 M KCl solution is 0.1411 S m^{-1}. All data are at 298.2 K.

Solution

Step 1 Clarify and define the problem

We want the molar conductivity $\Lambda(HA)$ of a weak acid HA in a solution.

Step 2 Select the key equation

We start, as recommended in Chapter 1, with the defining equation for the required quantity which, by the application of **Equation 11.3**, is

$$\Lambda(HA) = \frac{\kappa(HA)}{c(HA)} \qquad \text{(1) Equation 11.3}$$

Equation (1) shows that to calculate $\Lambda(HA)$ we need $\kappa(HA)$ and $c(HA)$ which are respectively the conductivity and the molar concentration of HA in the weak acid solution. Look at the data; $c(HA)$ is given while $\kappa(HA)$ is not. We must therefore try to see how $\kappa(HA)$ can be calculated from the data given.

Step 3 Derive the equation for the calculation

The conductivity $\kappa(HA)$ of the weak acid solution is given by **Equation 11.2** as

$$\kappa(HA) = \frac{G(HA)\, l}{a} \qquad \text{Equation 11.2}$$

$$= \frac{1}{R(HA)} \times \frac{l}{a} \qquad \text{(2) (by substituting Equation 11.1)}$$

In equation (2), l/a is unknown; it can, however, be calculated from the conductivity and resistance data given for the KCl solution. Application of **Equation 11.2** to the KCl solution shows that

$$\frac{l}{a} = \frac{\kappa(KCl)}{G(KCl)} \qquad \text{Equation 11.2}$$

$$= \kappa(KCl) \times R(KCl) \qquad (3)$$

On substituting for (l/a) in equation (2) by equation (3) we obtain

$$\kappa(HA) = \frac{1}{R(HA)} \times \kappa(KCl) \times R(KCl) \qquad (4)$$

Replacement of $\kappa(HA)$ in equation (1) by equation (4) then gives

$$\Lambda(HA) = \frac{\kappa(KCl)\, R(KCl)}{R(HA)\, c(HA)} \qquad (5)$$

Equation (5) can be used to calculate the required quantity $(\Lambda(HA))$ because all the quantities required for its calculation are given in the data.

Step 4 Collect the data, check the units, and calculate

$$\kappa(KCl) = 0.1411 \text{ S m}^{-1}$$
$$R(KCl) = 18.98 \ \Omega$$
$$R(HA) = 43.91 \ \Omega$$
$$c(HA) = 0.1000 \text{ mol dm}^{-3}$$
$$= 1.000 \times 10^2 \text{ mol m}^{-3}$$

$$\therefore \quad \Lambda(HA) = \frac{0.1411 \text{ S m}^{-1} \times 18.98 \ \Omega}{43.91 \ \Omega \times 1.000 \times 10^2 \text{ mol m}^{-3}}$$

$$= 6.099 \times 10^{-4} \text{ S m}^2 \text{ mol}^{-1}$$

The molar conductivity of the weak acid HA in the solution given is therefore 6.099 $\times 10^{-4}$ S m^2 mol^{-1}.

Step 5 Review and check the solution

In the calculation it was implicitly assumed that the conductivity of the electrolyte in the solution is equal to the conductivity of the solution. That is, the conductivity due to the solvent was assumed to be negligibly small. This assumption can be made except when the concentration of the electrolyte in the solution is very small (less than about 10^{-5} M).

Note

The molar conductivity of an electrolyte in a solution is normally determined from experimental measurements of resistance. The data required and the steps in the determination should be clear from Example 11.1.

Exercise 11.1

The resistivity (resistivity is defined as the reciprocal of conductivity) of a 0.100 M solution of a salt is $8.00 \times 10^{-1} \ \Omega$ m. Calculate the molar conductivity of the salt in the solution.

Exercise 11.2

The conductivity of a solution is 18.2 S m^{-1}. What potential difference (V) must be applied across two electrodes 15.0 cm apart and each of area 25.0 cm^2 so that the current (I) flowing is 1.00 A. Assume that Ohm's law ($I = V/R$) is applicable.

Exercise 11.3

Students often find it helpful to get a 'physical picture' or the 'physical significance' of the physical quantities that they have to study and understand.

(a) Analyse **Equation 11.2** and recognize that the conductivity (κ) may be imagined to be the conductance of a cube of solution of unit side.
(b) Show that the molar conductivity Λ of an electrolyte in a solution of volume V is related to the moles of electrolyte n present in the solution by the equation $\Lambda = \kappa V/n$. Hence recognize that the molar conductivity may be imagined to be the conductivity when one mole of solute is present in unit volume.

11.2 Conductivities of ions

Conductivity is found to be an *additive* property; the conductivity κ of a solution is the sum of the individual conductivities of all the ions present in the solution. That is

$$\kappa = \sum \kappa_i \qquad \textbf{Equation 11.4a}$$

where κ_i is the conductivity of an ion i in the solution.

The molar conductivity of an ion (Λ_i) in a solution is related to the conductivity κ_i of that ion by an equation which is exactly analogous to **Equation 11.3**. That is

$$\Lambda_i = \frac{\kappa_i}{c_i}$$

Here c_i refers to the concentration of the ion i in the solution.

Since conductivities are additive it follows (see Exercise 11.5) that the molar conductivity of an electrolyte (Λ) must be related to the molar conductivities of the ions (Λ_i) of the electrolyte by the equation

$$\Lambda = \sum v_i \Lambda_i \qquad \textbf{Equation 11.4b}$$

where v_i is the number of ions of type i formed from the electrolyte. For example, for iron(III) sulphate $(Fe_2(SO_4)_3)$, v_i would be 2 for the iron(III) ions and 3 for the sulphate ions.

Exercise 11.4

Recognize that the application of **Equation 11.4b** to aqueous solutions of Na_2SO_4 and $FeCl_3$ would give respectively the following equations:

$$\Lambda(Na_2SO_4) = 2\Lambda(Na^+) + \Lambda(SO_4^{2-})$$

$$\Lambda(FeCl_3) = \Lambda(Fe^{3+}) + 3\Lambda(Cl^-)$$

Exercise 11.5

Consider a 10.0 mol m^{-3} aqueous solution of Na_2SO_4. Apply **Equation 11.3** and write down the equations for the molar conductivities of Na_2SO_4, Na^+, and SO_4^{2-}. By making use of **Equation 11.4** then prove that

$$\Lambda(Na_2SO_4) = 2\Lambda(Na^+) + \Lambda(SO_4^{2-})$$

Example 11.2 Calculation of the conductivity of a solution containing several ions

Calculate the conductivity of the solution obtained by mixing 100 cm^3 of 0.0800 M HCl with 100 cm^3 of 0.0200 M NaCl. Assume that

$$\Lambda(Na^+) = \Lambda(Cl^-) = 6 \times 10^{-3} \text{ S m}^2 \text{ mol}^{-1}$$

$$\Lambda(H^+) = 35 \times 10^{-3} \text{ S m}^2 \text{ mol}^{-1}$$

Solution

Step 1 Clarify and define the problem

The total conductivity κ of the solution is required. The molar conductivity Λ_i of the ions present in the solution are given in the data. We must therefore try to relate κ to Λ_i values.

Step 2 Select the key equation

The conductivity κ of a solution is the sum of the conductivities (κ_i) of all the ions present in the solution. This statement could be given in equation form as

$$\kappa = \Sigma\, \kappa_i \qquad \text{(1)} \ \textbf{Equation 11.4a}$$

Step 3 Derive the equation for the calculation

In the solution considered the ions present are H^+, Na^+, and Cl^- (the amount of OH^- ions formed by the dissociation of the solvent water is relatively small and has therefore been neglected). Equation (1) could therefore be written more specifically as

$$\kappa = \kappa(H^+) + \kappa(Na^+) + \kappa(Cl^-) \qquad (2)$$

To calculate κ we must therefore know $\kappa(H^+)$, $\kappa(Na^+)$, and $\kappa(Cl^-)$ in the solution; these values are however not given in the data. Can we calculate these κ_i values from the data given? Yes, by the application of **Equation 11.3**. Application of this equation shows that

$$\kappa(H^+) = \Lambda(H^+) \times c(H^+) \qquad (3)$$

$$\kappa(Na^+) = \Lambda(Na^+) \times c(Na^+) \qquad (4)$$

$$\kappa(Cl^-) = \Lambda(Cl^-) \times c(Cl^-) \qquad (5)$$

Substitution of equations (3), (4), and (5) in equation (2) then shows that

$$\kappa = \Lambda(H^+)\, c(H^+) + \Lambda(Na^+)\, c(Na^+) + \Lambda(Cl^-)\, c(Cl^-) \qquad (3)$$

Step 4 Collect the data, check the units, and calculate

The data show that

$$\Lambda(H^+) = 35 \times 10^{-3}\ S\ m^2\ mol^{-1}$$

$$\Lambda(Na^+) = \Lambda(Cl^-) = 6 \times 10^{-3}\ S\ m^2\ mol^{-1}$$

Since the conductance data given are in metre (m) units it is clear that we must give the required ionic concentrations too in metre units (i.e. as moles per cubic metre, mol m^{-3}). Since HCl and NaCl are fully dissociated it is easy to show that (apply **Equation 4.2**, remembering that the total volume is now 200 cm³)

$$c(H^+) = 0.0400\ mol\ dm^{-3} = 40.0\ mol\ m^{-3}$$

$$c(Na^+) = 0.0100\ mol\ dm^{-3} = 10.0\ mol\ m^{-3}$$

$$c(Cl^-) = (0.0400 + 0.0100)\ mol\ cm^{-3} = 50.0\ mol\ m^{-3}$$

By the application of equation (6) we then obtain

$$\kappa = (35 \times 10^{-3} \text{ S m}^2 \text{ mol}^{-1} \times 40.0 \text{ mol m}^{-3})$$
$$+ (6 \times 10^{-3} \text{ S m}^2 \text{ mol}^{-1} \times 10.0 \text{ mol m}^{-3})$$
$$+ (6 \times 10^{-3} \text{ S m}^2 \text{ mol}^{-1} \times 50.0 \text{ mol m}^{-3})$$
$$= 1.40 \text{ S m}^{-1} + 0.060 \text{ S m}^{-1} + 0.30 \text{ S m}^{-1}$$
$$= 1.76 \text{ S m}^{-1}$$

The conductivity of the solution is therefore 1.76 S m^{-1}.

Step 5 Review and check the solution

Example 11.3 Determination of Λ^∞ for a weak electrolyte

The molar conductivities at infinite dilution of sodium ethanoate, hydrochloric acid, and sodium chloride are 0.917×10^{-2} S m^2 mol^{-1}, 4.255×10^{-2} S m^2 mol^{-1}, and 1.264×10^{-2} S m^2 mol^{-1} respectively at 25 °C. Calculate the molar conductivity of ethanoic acid at infinite dilution.

Solution

Step 1 Clarify and define the problem

The molar conductivity at infinite dilute of ethanoic acid—i.e. $\Lambda^\infty(CH_3COOH)$—is required. This can be calculated from the data given by making use of the principle of additivity of ionic conductivities.

Step 2 Select the key equation

We start with the required quantity—$\Lambda^\infty(CH_3COOH)$. From the principle of additivity of ionic conductivities it is clear that

$$\Lambda^\infty(CH_3COOH) = \Lambda^\infty(CH_3COO^-) + \Lambda^\infty(H^+) \quad (1) \text{ Equation 11.4b}$$

Step 3 Derive the equation for the calculation

To calculate $\Lambda^\infty(CH_3COOH)$ by equation (1) we require $\Lambda^\infty(CH_3COO^-)$ and $\Lambda^\infty(H^+)$. These quantities can be related to the given data by the equations

$$\Lambda^\infty(CH_3COO^-) = \Lambda^\infty(CH_3COONa) - \Lambda^\infty(Na^+) \quad (2) \text{ Equation 11.4b}$$
$$\Lambda^\infty(H^+) = \Lambda^\infty(HCl) - \Lambda^\infty(Cl^-) \quad (3) \text{ Equation 11.4b}$$

On replacing $\Lambda^\infty(CH_3COO^-)$ and $\Lambda^\infty(H^+)$ in equation (1) by equations (2) and (3) we obtain

$$\Lambda^\infty(CH_3COOH) = \Lambda^\infty(CH_3COONa) - \Lambda^\infty(Na^+) + \Lambda^\infty(HCl) - \Lambda^\infty(Cl^-)$$
$$= \Lambda^\infty(CH_3COONa) + \Lambda^\infty(HCl) - (\Lambda^\infty(Na^+) + \Lambda^\infty(Cl^-))$$
$$= \Lambda^\infty(CH_3COONa) + \Lambda^\infty(HCl) - \Lambda^\infty(NaCl) \quad (4)$$

Equation (4) can be used to calculate the required quantity, $\Lambda^\infty(CH_3COOH)$.

Step 4 Calculate the data, check the units, and calculate

$$\Lambda^\infty(CH_3COONa) = 0.917 \times 10^{-2} \text{ S m}^2 \text{ mol}^{-1}$$

$$\Lambda^\infty(HCl) = 4.255 \times 10^{-2} \text{ S m}^2 \text{ mol}^{-1}$$

$$\Lambda^\infty(NaCl) = 1.264 \times 10^{-2} \text{ S m}^2 \text{ mol}^{-1}$$

$$\therefore \quad \Lambda^\infty(CH_3COOH) = (0.917 \times 10^{-2} + 4.255 \times 10^{-2} - 1.264 \times 10^{-2}) \text{ S m}^2 \text{ mol}^{-1}$$

$$= 3.908 \times 10^{-2} \text{ S m}^2 \text{ mol}^{-1}$$

The molar conductivity of ethanoic acid at infinite dilution is therefore 3.908×10^{-2} S m^2 mol^{-1}.

Step 5 Review and check the solution

Note

This example illustrates one method commonly used for obtaining the molar conductivity at infinite dilution (Λ^∞) of weak electrolytes. For weak electrolytes Λ^∞ cannot be obtained directly by experiment since such electrolytes do not dissociate completely in solution even at high dilutions.

The principle involved in this calculation is that the conductivity of an ion depends only on that ion and is independent of the other ions present in the solution (this principle is strictly true only at infinite dilution). We may therefore add and subtract known conductivities of electrolytes so as to obtain the conductivity of another electrolyte.

Exercise 11.6

The molar conductivity at infinite dilution (Λ^∞) of a solution of $(NH_4)_2SO_4$ is 2.33×10^{-2} S m^2 mol^{-1}. Λ^∞ for NH_4^+ ion is 7.45×10^{-3} S m^2 mol^{-1}. Calculate Λ^∞ for SO_4^{2-} ions.

Exercise 11.7

The molar conductivity of a 0.0100 M solution of a weak acid HA is 1.20×10^{-2} S m^2 mol^{-1}. If the molar conductivities of H^+ and A^- ions are 3.50×10^{-2} S m^2 mol^{-1} and 2.00×10^{-2} S m^2 mol^{-1} respectively, calculate the degree of dissociation of HA in the solution.

Exercise 11.8

Λ^∞ for NH_4Cl, NaCl, and NaOH are 1.50×10^{-2} S m^2 mol^{-1}, 1.35×10^{-2} S m^2 mol^{-1}, and 2.58×10^{-2} S m^2 mol^{-1} respectively. Calculate Λ^∞ for NH_4OH.

Exercise 11.9

A solution contains 0.500 M Ca^{2+} ions, 0.200 M H^+ ions, and 1.00 M Cl^- ions. Calculate the conductivity of the solution if the molar conductivities of the Ca^{2+}, H^+, and Cl^- ions in the solution are 1.19×10^{-2} S m^2 mol^{-1}, 3.50×10^{-2} S m^2 mol^{-1}, and 7.44×10^{-3} S m^2 mol^{-1} respectively.

11.3 Transference number

The transference number t_i of an ion i in a solution is defined as the fraction of the conductivity due to that ion. That is

$$t_i = \frac{\kappa_i}{\kappa} \qquad \text{Equation 11.5}$$

where κ_i is the conductivity due to the ion i and κ is the total conductivity due to all the ions in the solution. (Transference numbers were formerly known as transport numbers.)

Example 11.4 Molar conductivities of ions from transference numbers

A conductivity cell containing 0.100 M Na_2SO_4 has a resistance 20.1 Ω. The cell constant (l/a) is 0.520 cm^{-1}. If the transference number of the Na$^+$ ion in this solution is 0.385, calculate the molar conductivity in the solution of (a) Na$^+$ ions (b) SO_4^{2-} ions.

Solution to Part (a)

Step 1 Clarify and define the problem

We have to calculate the molar conductivity of the Na$^+$ ion, $\Lambda(Na^+)$, from its transference number, $t(Na^+)$, in the solution. To do so we must derive an equation that shows the relationship between these two quantities.

Step 2 Select the key equation

We start with the defining equation for $\Lambda(Na^+)$. Application of this equation shows that

$$\Lambda(Na^+) = \frac{\kappa(Na^+)}{c(Na^+)} \qquad (1)\ \text{Equation 11.3}$$

Step 3 Derive the equation for the calculation

$c(Na^+)$ is seen from the data to be 0.200 M and therefore to calculate $\Lambda(Na^+)$ by equation (1) we must be able to find out $\kappa(Na^+)$ from the data given. $\kappa(Na^+)$ is related to the data given $(t(Na^+))$ by the equation

$$\kappa(Na^+) = t(Na^+) \times \kappa \qquad (2)\ \text{Equation 11.5}$$

On substituting for $\kappa(Na^+)$ in equation (1) by equation (2) we get

$$\Lambda(Na^+) = \frac{t(Na^+) \times \kappa}{c(Na^+)} \qquad (3)$$

κ is unknown; it is, however, related to the data given by the equation (use **Equations 11.2 and 11.1**)

$$\kappa = \frac{1}{R} \times \frac{l}{a} \qquad (4)$$

Equation (3) can, therefore, be rewritten as

$$\Lambda(Na^+) = \frac{t(Na^+)}{c(Na^+)} \times \frac{l}{Ra} \tag{5}$$

Step 4 Collect the data, check the units, and calculate

$$t(Na^+) = 0.385$$
$$(l/a) = 0.520 \text{ cm}^{-1} = 0.520 \times 10^2 \text{ m}^{-1}$$
$$c(Na^+) = 0.200 \text{ M} = 2.00 \times 10^2 \text{ mol m}^{-3}$$
$$R = 20.1 \ \Omega$$

$$\therefore \quad \Lambda(Na^+) = \frac{0.385 \ (0.520 \times 10^2 \text{ m}^{-1})}{(2.00 \times 10^2 \text{ mol m}^{-3}) \ (20.1 \ \Omega)}$$

$$= 4.97 \times 10^{-3} \text{ S m}^2 \text{ mol}^{-1}$$

The molar conductivity of the Na^+ ion is therefore 4.97×10^{-3} S m² mol⁻¹.

Solution to Part (b)

$\Lambda(SO_4^{2-})$ could be calculated similarly. For this calculation (using an equation analogous to equation 5) we require $t(SO_4^{2-})$. This is easily calculated. From the definition of transference number it follows that the sum of the transference numbers of all the ions in a solution must be equal to unity. In a Na_2SO_4 solution it therefore follows that

$$t(Na^+) + t(SO_4^{2-}) = 1$$
$$\therefore \quad t(SO_4^{2-}) = 1 - t(Na^+)$$
$$= 0.615$$

Calculation using an equation analogous to equation 5 shows that

$$\Lambda(SO_4^{2-}) = 1.59 \times 10^{-2} \text{ S m}^2 \text{ mol}^{-1}$$

Step 5 Review and check the solution

Exercise 11.10

Consider an aqueous solution of $Al_2(SO_4)_3$. Derive the equations that show the relationship between the molar conductivities of the Al^{3+} ions and SO_4^{2-} ions and (a) the transference number of the Al^{3+} ion (b) the transference number of the SO_4^{2-} ion.

Exercise 11.11

Calculate the transference number of the Cl^- ion in an aqueous solution of iron(III) chloride $FeCl_3$. The molar conductivities of the Fe^{3+} and Cl^- ions in the solution are 1.6×10^{-2} S m² mol⁻¹ and 7.5×10^{-3} S m² mol⁻¹ respectively.

11.4   Applications of conductance measurements

Many of the applications of conductance measurements are based on the determination of the concentrations of the *ions* in a solution: this determination is based on **Equation 11.3** which for a particular ion i can be written as

$$\Lambda_i = \frac{\kappa_i}{c_i}$$
Equation 11.3

Two applications are illustrated in Examples 11.5 and 11.6.

Example 11.5   Determination of dissociation constant from conductance data

The molar conductivity of a 0.100 M solution of a weak acid HA at 25 °C is 5.17×10^{-4} S m^2 mol^{-1}. At this temperature Λ^∞ for H^+ and A^- ions are 3.50×10^{-2} S m^2 mol^{-1} and 0.40×10^{-2} S m^2 mol^{-1} respectively. Calculate (a) the degree of dissociation of HA (b) the dissociation constant of HA.

Solution to Part (a)

Step 1   Clarify and define the problem

Let α be the required degree of dissociation. To calculate α from the Λ data given it is necessary to derive an equation showing the relation between α and Λ. To do so some clarification of the problem is necessary; it would also be necessary to make certain assumptions.

Conductance is due only to the ions present; Λ for an electrolyte would depend, therefore, on the amount of ions formed from the electrolyte.

Step 2   Select the key equation

In a solution of a weak electrolyte HA only a fraction α of the electrolyte is dissociated into ions (α = fraction of dissociation or degree of dissociation). It therefore appears reasonable to assume that Λ would be directly proportional to α. That is

$$\Lambda \propto \alpha \tag{1}$$

At infinite dilution the electrolyte is always fully dissociated. That is α = 1. Equation (1) would then become

$$\Lambda^\infty \propto 1 \tag{2}$$

On dividing equation (1) by (2) we obtain

$$\frac{\Lambda}{\Lambda^\infty} = \alpha \tag{3}$$

Step 3   Derive the equation for the calculation

Equation (3) shows that the required quantity α could be calculated if Λ and Λ^∞ are known. Λ is given in the data. The principle of additivity of ionic conductivities shows that Λ^∞ of HA is given by

$$\Lambda^\infty = \Lambda^\infty(H^+) + \Lambda^\infty(A^-) \tag{4 Equation 11.4b}$$

From equations (3) and (4) it then follows that

$$\alpha = \frac{\varLambda}{\varLambda^{\infty}(\text{H}^+) + \varLambda^{\infty}(\text{A}^-)} \qquad (5)$$

Step 4   Collect the data, check the units, and calculate

$$\varLambda = 5.17 \times 10^{-4}\,\text{S m}^2\,\text{mol}^{-1}$$

$$\varLambda^{\infty}(\text{H}^+) = 3.50 \times 10^{-2}\,\text{S m}^2\,\text{mol}^{-1}$$

$$\varLambda^{\infty}(\text{A}^-) = 0.40 \times 10^{-2}\,\text{S m}^2\,\text{mol}^{-1}$$

$$\therefore \qquad \alpha = \frac{5.17 \times 10^{-4}\,\text{S m}^2\,\text{mol}^{-1}}{(3.50 \times 10^{-2} + 0.40 \times 10^{-2})\,\text{S m}^2\,\text{mol}^{-1}}$$

$$= 1.33 \times 10^{-2}$$

The degree of dissociation of HA in the given solution is therefore 1.33×10^{-2}.

Solution to Part (b)

Step 1   Clarify and define the problem

Since α has been calculated in Part (a), let us see how the required quantity—the dissociation constant K—can be calculated from α. To do so we have to derive an equation which relates K and α.

Step 3   Select the key equation

We focus our attention on the required quantity—K for the dissociation of HA. This is given by the equation

$$K = \frac{[\text{H}^+]\,[\text{A}^-]}{[\text{HA}]} \qquad \text{(6) Equation 7.1}$$

Step 3   Derive the equation for the calculation

To calculate K by equation (6), $[\text{H}^+]$, $[\text{A}^-]$, and $[\text{HA}]$ are required; these can be related to α as follows:

From the definition of α (see **Equation 2.2**) it is clear that

$$\alpha = \frac{\text{moles of HA dissociated}}{\text{moles of HA present initially}} \qquad (7)$$

'Moles' in equation (7) can be replaced by 'concentration' because **Equation 4.2** ($c = n/V$) shows that concentration (c) is directly proportional to the number of moles (n). Equation (7) can therefore be rewritten as

$$\alpha = \frac{\text{concentration of HA dissociated}}{\text{initial concentration of HA}} \qquad (8)$$

The balanced equation for the dissociation

$$HA(aq) = H^+(aq) + A^-(aq)$$

shows that

concentration of HA dissociated = concentration of H^+ ions
formed by dissociation

$$= [H^+] \tag{9}$$

On inserting equation (9) in equation (8) and on rearranging we obtain

$$[H^+] = \alpha \times c \tag{10}$$

where $c =$ initial concentration of HA. Equation (10) relates $[H^+]$ to α.

We have still to relate $[A^-]$ and $[HA]$ to α. This can be done with the help of the balanced equation for the dissociation which shows that

$$[A^-] = [H^+]$$
$$= \alpha \times c \tag{11}$$
$$[HA] = c - [H^+]$$
$$= c - (\alpha \times c)$$
$$= c (1 - \alpha) \tag{12}$$

We can now obtain the required equation for the calculation by replacing $[H^+]$, $[A^-]$, and $[HA]$ in equation (6) by equations (10), (11), and (12).

$$K = \frac{\alpha c \times \alpha c}{c (1 - \alpha)}$$

$$= \frac{\alpha^2 c}{1 - \alpha} \tag{13}$$

Step 4 Collect the data, check the units, and calculate

$$\alpha = 1.33 \times 10^{-2} \qquad \text{(from solution (a))}$$

$$c = 0.100 \text{ mol dm}^{-3} \qquad \text{(data)}$$

$$\therefore \qquad K = \frac{(1.33 \times 10^{-2})^2 \times 0.100 \text{ mol dm}^{-3}}{(1 - 0.0133)}$$

$$= 1.76 \times 10^{-5} \text{ mol dm}^{-3}$$

The dissociation constant of the weak acid HA is therefore 1.76×10^{-5} mol dm^{-3}.

Step 5 Review and check the solution

Example 11.6 Determination of solubility product from conductivity data

The conductivity of a saturated aqueous solution of silver chloride is 1.980×10^{-4} S m^{-1}. If the conductivity of the water used to prepare the solution is 1.78×10^{-5} S m^{-1}, calculate (a) the solubility (b) the solubility product of silver chloride. Λ^∞ for Ag$^+$ and Cl$^-$ are 6.35×10^{-3} S m^2 mol^{-1} and 7.55×10^{-3} S m^2 mol^{-1} respectively. All data are at 25 °C.

Solution to Part (a)

Step 1 Clarify and define the problem

The term solubility refers to the concentration of the solute in a saturated solution. Let c be the required quantity—the solubility of silver chloride in water.

Step 2 Select the key equation

To calculate c from the conductivity data given we make use of the equation

$$c = \frac{\kappa}{\Lambda} \qquad \text{(1) Equation 11.3}$$

In this equation κ and Λ are respectively the conductivity and molar conductivity due only to the *solute* in the solution.

Step 3 Derive the equation for the calculation

The experimentally measured conductivity of a saturated solution of AgCl is due to the solute (Ag$^+$ and Cl$^-$ ions) and the solvent (H$^+$ and OH$^-$ from H$_2$O). It therefore follows that the conductivity due only to the solute (κ) in a solution is given by

$$\kappa = \kappa_{\text{solution}} - \kappa_{\text{solvent}} \qquad (2)$$

Consider now Λ in equation (1). We may assume that Λ is equal to Λ^∞ (see note 2 below) and therefore write

$$\Lambda = \Lambda^\infty(\text{Ag}^+) + \Lambda^\infty(\text{Cl}^-) \qquad \text{(3) Equation 11.4b}$$

On inserting equations (2) and (3) in equation (1) we obtain

$$c = \frac{\kappa_{\text{solution}} - \kappa_{\text{solvent}}}{\Lambda^\infty(\text{Ag}^+) + \Lambda^\infty(\text{Cl}^-)} \qquad (4)$$

Step 4 Collect the data, check the units, and calculate

$$\kappa_{\text{solution}} = 1.980 \times 10^{-4} \text{ S m}^{-1}$$

$$\kappa_{\text{solvent}} = 1.78 \times 10^{-5} \text{ S m}^{-1} = 0.178 \times 10^{-4} \text{ S m}^{-1}$$

$$\Lambda^\infty(\text{Ag}^+) = 6.35 \times 10^{-3} \text{ S m}^2 \text{ mol}^{-1}$$

$$\Lambda^\infty(\text{Cl}^-) = 7.55 \times 10^{-3} \text{ S m}^2 \text{ mol}^{-1}$$

$$\therefore \quad c = \frac{1.980 \times 10^{-4} \text{ S m}^{-1} - 0.178 \times 10^{-4} \text{ S m}^{-1}}{6.35 \times 10^{-3} \text{ S m}^2 \text{ mol}^{-1} + 7.55 \times 10^{-3} \text{ S m}^2 \text{ mol}^{-1}}$$

$$= 1.296 \times 10^{-2} \text{ mol m}^{-3}$$

The solubility of AgCl in water at 25 °C is therefore 1.296×10^{-2} mol m^{-3}, that is 1.30×10^{-5} mol dm^{-3}.

Solution to Part (*b*)

The solubility product K of AgCl is required. It is defined by the equation

$$K = [Ag^+][Cl^-] \qquad \text{(see section 8.5)}$$

In a solution of silver chloride $[Ag^+]$ is equal to $[Cl^-]$ because 1 mole of AgCl produces 1 mole of Ag^+ and 1 mole of Cl^- on dissociation. Also the concentration of each ion in the solution is equal to the solubility of AgCl since there is complete dissociation. Therefore

$$K = [Ag^+][Cl^-]$$
$$= (1.30 \times 10^{-5} \text{ mol dm}^{-3})^2$$
$$= 1.69 \times 10^{-10} \text{ mol}^2 \text{ dm}^{-6}$$

Step 5   Review and check the solution

Notes

1 Since the solubility of AgCl is very small, the conductivity due to the solute in the solution is also very small. The conductivity of the solvent would not then be negligibly small compared with the conductivity of the solute. A correction must therefore be made for the conductivity of the solvent (this was done in the calculation above). If the conductivity due to the solute is high, the conductivity of the solvent may be neglected (as was done in Example 11.5).

2 Since the solubility of AgCl is small, the concentrations of Ag^+ and Cl^- ions in a saturated solution of AgCl are also small. The molar conductivity of AgCl in a saturated solution may therefore be assumed to be equal to that at infinite dilution. This assumption was made during the calculation.

Exercise 11.12

The resistance of a 0.100 M solution of a weak acid HA in a conductivity cell of cell constant $2.00 \times 10^2 \text{ m}^{-1}$ is 439.1 Ω. Calculate the dissociation constant of the acid if Λ_i^∞ for the H^+ and A^- ions are $3.50 \times 10^{-2} \text{ S m}^2 \text{ mol}^{-1}$ and $5.20 \times 10^{-3} \text{ S m}^2 \text{ mol}^{-1}$ respectively.

Exercise 11.13

Calculate the conductivity of Ag_2CrO_4 in a saturated solution of Ag_2CrO_4 if the solubility product of Ag_2CrO_4 is $9.0 \times 10^{-12} \text{ mol}^3 \text{ dm}^{-9}$. The conductivity of the water used to prepare the solution is $1.0 \times 10^{-5} \text{ S m}^{-1}$ and Λ_i for Ag^+ and CrO_4^{2-} are $6 \cdot 35 \times 10^{-3} \text{ S m}^2 \text{ Mol}^{-1}$ and $1.11 \times 10^{-2} \text{ S m}^2 \text{ mol}^{-1}$ respectively. Remember that since the concentrations of Ag^+ and CrO_4^{2-} in the saturated soution are small, the value of Λ in the solution may be taken as equal to Λ^∞.

Exercise 11.14

The conductivity of pure water is $5.53 \times 10^{-6} \text{ S m}^{-1}$ at 25 °C. Calculate the ionic product of water. Λ_i^∞ for H^+ and OH^- at 25 °C are $3.48 \times 10^{-2} \text{ S m}^2 \text{ mol}^{-1}$ and $1.98 \times 10^{-2} \text{ S m}^2 \text{ mol}^{-1}$ respectively.

11.5 Conductance titrations

Conductance measurements are also often useful for finding the equivalence point of a titration. This method is possible whenever there is an *abrupt change* in the conductance of the solution at the equivalence point. The abrupt change may be caused by a change either in the nature of the ions or in the concentrations of the ions.

A plot of the variation of the conductance of the solution with the volume of titrant added is known as a conductance titration graph.

Example 11.7 Conductance titration

100 cm³ of a 0.0500 M solution of HCl, in a conductivity cell, is conductometrically titrated with 1.00 M NaOH solution. Calculate the conductivity of the solution after the addition of 0, 1.00, 3.00, 5.00, 8.00, and 10.0 cm³ of NaOH. Assume that all univalent ions, except H^+ and OH^-, have a molar conductivity of 6×10^{-3} S m² mol⁻¹. $\Lambda(H^+) = 35 \times 10^{-3}$ S m² mol⁻¹ and $\Lambda(OH^-) = 20 \times 10^{-3}$ S m² mol⁻¹. Assume that molar conductivities of the ions do not change with concentration. Neglect the volume change on the addition of NaOH.

Plot the conductance titration graph.

Solution

Step 1 Clarify and define the problem

We have to calculate the conductivity κ of a solution. The molar conductivities (Λ_i) of the ions present in the solution are given in the data. We have therefore to derive an equation which relates κ to Λ_i values.

Step 2 Select the key equation

The conductivity κ of a solution is the sum of the conductivities of all the ions in the solution (see **Equation 11.4a**). In the example considered H^+, Cl^-, Na^+, and OH^- are present in the solution and therefore

$$\kappa = \kappa(H^+) + \kappa(Cl^-) + \kappa(Na^+) + \kappa(OH^-) \quad \text{(1) Equation 11.4a}$$

Step 3 Derive the equation for the calculation

To calculate κ we must therefore be able to calculate κ_i for each of the ions in the solution. κ_i values could be calculated by making use of **Equation 11.3**. For an ion i this equation is

$$\kappa_i = c_i \Lambda_i \quad \text{(2) Equation 11.3}$$

On replacing each of the κ_i terms given in equation (1) by the application of equation (2) we obtain

$$\kappa = c(H^+) \Lambda(H^+) + c(Cl^-) \Lambda(Cl^-) + c(Na^+) \Lambda(Na^+) + c(OH^-) \Lambda(OH^-) \quad \text{(3)}$$

To calculate κ using equation (3) we must therefore first calculate the concentration of each ion (c_i) in the solution. This could be done by making use of the balanced equation (see Chapter 4) for the titration reaction which is

$$OH^- \text{ (from NaOH)} + H^+ \text{ (from HCl)} \rightarrow H_2O$$

Step 4 Collect the data, check the units, and calculate

The values of c_i calculated by making use of the balanced equation are given in rows 1–4 of Table 11.1. The next four rows 5–8 show the κ_i values calculated using equation (2). The final row 9 indicates the required total conductivity κ of the solution, calculated using equation (3), after each addition of NaOH.

Table 11.1   Data calculated in Example 11.7

cm^3 of NaOH added	0.0	1.0	3.0	5.0	8.0	10.0
1 $c(H^+)/mol\ m^{-3}$	50.0	40.0	20.0	—	—	—
2 $c(Cl^-)/mol\ m^{-3}$	50.0	50.0	50.0	50.0	50.0	50.0
3 $c(Na^+)/mol\ m^{-3}$	—	10.0	30.0	50.0	80.0	100.0
4 $c(OH^-)/mol\ m^{-3}$	—	—	—	—	30.0	50.0
5 $\kappa(H^+)/S\ m^{-1}$	1.75	1.40	0.70	—	—	—
6 $\kappa(Cl^-)/S\ m^{-1}$	0.30	0.30	0.30	0.30	0.30	0.30
7 $\kappa(Na^+)/S\ m^{-1}$	—	0.06	0.18	0.30	0.48	0.60
8 $\kappa(OH^-)/S\ m^{-1}$	—	—	—	—	0.60	1.00
9 $\therefore \kappa/S\ m^{-1}$	2.05	1.76	1.18	0.60	1.38	1.90

Step 5 Review and check the solution

To obtain the conductance titration graph we have to plot the conductance (in any units since the units do not matter) on the y axis against the volume of sodium hydroxide added (on x axis). The plot of the results given in Table 11.1 is shown in Fig. 11.1.

In this particular case, the minimum conductance is seen to correspond to the equivalence point. For other conductance titrations the shape of the graph may be different.

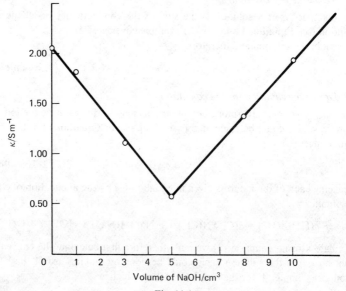

Fig. 11.1

Note

To be able to deduce the shapes of the various conductance titration graphs, you should remember that the molar conductivities of H^+ ions (35×10^{-3} S m^2 mol^{-1} at 25 °C) and OH^- ions (20×10^{-3} S m^2 mol^{-1} at 25 °C) are much larger than those for other ions.

Exercise 11.15

Look at the results given in Table 11.1 (Example 11.7) and explain

    (a) why the conductivity decreases until the equivalence point
    (b) why the conductivity increases after the equivalence point.

Exercise 11.16

Calculate the approximate conductivities of the solutions obtained when (a) 1.00 cm^3 (b) 2.00 cm^3 (c) 2.50 cm^3 (d) 3.00 cm^3 (e) 4.00 cm^3 (f) 5.00 cm^3, of 1.00 M AgNO$_3$ solution are added to 25.0 cm^3 of a 0.100 M solution of sodium chloride. Remember that all ions except H^+ and OH^- have approximate molar conductivity of $z_i \times 6 \times 10^{-3}$ S m^2 mol^{-1} where $z_i =$ change on the ion. Plot the conductance titration curve and recognize that the equivalence point of precipitation titrations can also be obtained from conductance measurements.

Exercise 11.17

Calculate the approximate conductivities of the solutions obtained when (a) 1.00 cm^3 (b) 2.00 cm^3 (c) 2.50 cm^3 (d) 3.00 cm^3 (e) 4.00 cm^3 (f) 5.00 cm^3, of 1.00 M solution of hydrochloric acid are added to 25.0 cm^3 of a 0.100 M solution of sodium hydroxide. Plot the conductance titration curve. $\Lambda(H^+) = 3.5 \times 10^{-2}$ S m^2 mol^{-1}, $\Lambda(OH^-) = 2.0 \times 10^{-2}$ S m^2 mol^{-1}. For all other ions $\Lambda_i = z_i \times 6 \times 10^{-3}$ S m^2 mol^{-1}, where z_i is the charge number of the ion.

Review of Chapter 11

Conductance G is the reciprocal of resistance R

$$G = \frac{1}{R} \qquad\qquad \textbf{11.1}$$

The *conductivity* of a solution is a characteristic property of the solution; it is the constant of proportionality κ in the equation:

$$G = \kappa \frac{a}{l} \qquad\qquad \textbf{11.2}$$

where a is the cross section area of the electrodes and l is the distance between them.

The *molar conductivity* (Λ) is defined as the conductivity (κ) per unit concentration. That is

$$\Lambda = \frac{\kappa}{c} \qquad\qquad \textbf{11.3}$$

This equation, which is applicable both to an individual ion as well as to an electrolyte, is important because many of the applications of conductance measurements relate to calculations of concentration c.

The conductivity, κ, of a solution is the sum of the individual conductivities of all the ions (κ_i) in the solution; that is

$$\kappa = \sum \kappa_i \qquad \text{11.4a}$$

There is a corresponding equation for molar conductivities

$$\Lambda = \sum v_i \Lambda_i \qquad \text{11.4b}$$

The fraction of the conductivity due to an ion i in a solution is known as the transference number t_i of that ion. That is

$$t_i = \frac{\kappa_i}{\kappa} \qquad \text{11.5}$$

12 Chemical kinetics

12.1 First order reactions

A first order reaction is defined as one in which the rate is found to be directly proportional to one concentration term.

To represent this statement as an equation we have first to give the necessary symbols. Suppose that the first order reaction considered is

$$A \rightarrow products$$

The defining statement then indicates that the rate of this reaction can be given as

$$rate \propto c_A$$
$$= kc_A \tag{1}$$

where k is a constant.

In equation (1) we have still to express 'rate' in mathematical notation; this can be done in various ways. Because c_A appears in the equation, it would be convenient to define the rate in terms of c_A. Since c_A decreases with time, as reaction proceeds, it would also be convenient to give the rate as the decrease in the concentration of A ($-dc_A$) divided by the corresponding time interval (dt). That is

$$rate = -\frac{dc_A}{dt}$$

Equation (1) could therefore be rewritten as

$$-\frac{dc_A}{dt} = kc_A \qquad \textbf{Equation 12.1a}$$

Equation 12.1a is the required mathematical expression for the rate of a first order reaction; k is a constant and is known as the rate constant.

In studies of reaction rates the quantities measured experimentally are the concentrations (c) at various times (t). **Equation 12.1a**, however, does not give the relation between c and t; instead it gives the relation between dc and dt. To test whether the experimental data for a reaction obeys **Equation 12.1a** it is therefore necessary to derive from **Equation 12.1a** an equation which shows the relation between c and t.

To obtain the relation between c and t, **Equation 12.1a** must be integrated. We now proceed to do so. On rearrangement **Equation 12.1a** becomes

$$\frac{dc}{c} = -k \, dt$$

which on integration, because

$$\int \frac{dc}{c} = \log_e c, \text{ and } \int dt = t$$

gives

$$\log_e c = -kt + I \tag{2}$$

where I is the integration constant. The value of the integration constant can be obtained by analysing equation (2) as follows. When $t = 0$, the concentration is evidently the initial concentration (say c_0). When $t = 0$, the term kt is also zero. If we make these two substitutions in equation (2) (i.e. put $c = c_0$ and $kt = 0$) and rearrrange we obtain

$$I = \log_e c_0 \tag{3}$$

If equation (3) is substituted in equation (2) we obtain

$$\log_e c = -kt + \log_e c_0$$

$$\therefore \qquad 2.303 \log_{10} c = -kt + 2.303 \log_{10} c_0 \qquad \text{(see **Equation 2.4**)}$$

$$\therefore \qquad \log_{10} c = -\frac{kt}{2.303} + \log_{10} c_0 \qquad \text{**Equation 12.1b**}$$

Equation 12.1b gives the required general relationship between c and t for a first order reaction.

Exercise 12.1

Analyse **Equation 12.1a**. Recognize that the rate constant could be considered to be the rate of the reaction when $c_A = 1$.

Exercise 12.2

Look at **Equation 12.1b**. State the data required for the calculation of the rate constant of a first order reaction.

Example 12.1 Rate constant of a first order reaction

A 1.000 M aqueous solution of a methyl ethanoate is hydrolysed at 298 K in the presence of acid (which acts as a catalyst). The equation for the hydrolysis is

$$CH_3COOCH_3 + H_2O \rightarrow CH_3COOH + CH_3OH$$

The concentration of ethanoic acid formed (which can be determined by titration with a strong alkali) as a function of time is given in Table 12.1.

Table 12.1 Concentration of ethanoic acid formed as a function of time in Example 12.1

time/minute	0	10.00	20.00	30.00	40.00
$[CH_3COOH]$/mol dm^{-3}	0	0.177	0.319	0.440	0.538

Calculate the rate constant at 298 K if the hydrolysis is first order.

Solution

Step 1 Clarify and define the problem

To calculate the rate constant k of a first order reaction we have to make use of **Equation 12.1b**. From this equation it is seen that k can be calculated if the initial concentration of the reactant c_0 is known and one experimental result of c (at a particular time t) is available.

Data give four results. Instead of calculating k for each result and then averaging, it is more convenient to make an appropriate plot of the results and obtain the required k value from the graph.

Step 2 Select the key equation

The rate constant k of a first order reaction is required. It is defined by the equation

$$\log_{10} c = -\frac{kt}{2.303} + \log_{10} c_0 \qquad \text{(1) Equation 12.1b}$$

where c_0 is the initial concentration of CH_3COOCH_3 and c is the concentration of CH_3COOCH_3 at a time t.

Equation (1) is of the form $y = mx + c$ if we take $\log_{10} c$ as y and t as x. A plot of $\log_{10} c$ (on the y-axis) against t (x-axis) would then give a straight line graph with a gradient equal to $(-k/2.303)$. From the gradient k can then be calculated.

Step 3 Derive the equation for the calculation

To plot $\log_{10} c$ vs t we must evidently know the c values; these are not given in the data. Data gives the concentration of ethanoic acid, $[CH_3COOH]$, in the solution. c values can, however, be calculated from $[CH_3COOH]$ and c_0 by making use of the equation for the reaction

$$CH_3COOCH_3 + H_2O \rightarrow CH_3COOH + CH_3OH$$

From the balanced equation it is easily seen that

$$c = c_0 - [CH_3COOH] \qquad (2)$$

Step 4 Collect the data, check the units, and calculate

$$c_0 = 1.000 \text{ mol dm}^{-3}$$

Values of c at various times t, calculated by equation (2), are given in the third row of Table 12.2 overleaf.

Table 12.2   Concentration of methyl ethanoate formed as a function
of time in Example 12.1

t/minute	0	10.00	20.00	30.00	40.00
$[CH_3COOH]$/mol dm^{-3}	0	0.177	0.319	0.440	0.538
c/mol dm^{-3}	1.000	0.823	0.681	0.560	0.462

If a plot of $\log_{10} c$ vs t is made the gradient of this graph, as seen from equation (1), will be given by

$$\text{gradient} = -\frac{k}{2.303}$$

$$\therefore \quad k = -\frac{\text{gradient}}{2.303}$$

$$= 1.92 \times 10^{-2} \text{ minute}^{-1} \qquad \text{(from the graph)}$$

$$= 3.20 \times 10^{-4} \text{ s}^{-1}$$

The rate constant for the given reaction is therefore 3.20×10^{-4} s^{-1}.

Step 5   Review and check the solution

Notes

1 The SI unit for the rate constant of a first order reaction is s^{-1}; it does not involve concentrations.

2 All calculations on first order reactions are based on **Equation 12.1b**. Study and analyse this equation carefully (see section 2.2 for the method of analysing equations).

Example 12.2 Calculation of amount of product from rate constant

The decomposition of benzenediazonium chloride ($C_6H_5N_2Cl$) in an acidic aqueous medium, is found to be a first order reaction. The equation for the decomposition is

$$C_6H_5N_2Cl + H_2O \rightarrow C_6H_5OH + HCl + N_2$$

At 300 K if the rate constant is 1.52×10^{-3} s^{-1}, calculate the moles of nitrogen molecules that would be evolved in 1.00 hour from 0.0500 dm^3 of a 1.000 M benzenediazonium chloride.

Solution

Step 1   Clarify and define the problem

Let the required quantity—the moles of nitrogen molecules evolved from 0.0500 dm^3 of a 1.000 mol dm^{-3} solution of $C_6H_5N_2Cl$ in 1.00 hour be denoted by n_{N_2}.

Step 2 Select the key equation

To calculate the required quantity—n_{N_2}—we start by making use of the information provided by the balanced equation for the reaction. From the balanced equation it is clear that

$$n_{N_2} = \text{moles of } C_6H_5N_2Cl \text{ decomposed} \qquad (1)$$

Step 3 Derive the equation for the calculation

Equation (1) can be simplified and rewritten as

$$n_{N_2} = \text{moles of } C_6H_5N_2Cl \text{ present initially}$$
$$- \text{ moles of } C_6H_5N_2Cl \text{ present after decomposition} \qquad (2)$$

By the application of **Equation 4.2** it is seen that

$$\text{mol } C_6H_5N_2Cl \text{ present initially} = c_0 \times V \qquad (3)$$

$$\text{mol } C_6H_5N_2Cl \text{ present after decomposition} = c \times V \qquad (4)$$

where (see data) $c_0 = 1.000$ mol dm^{-3}, $V = 0.0500$ dm^3, and c is the concentration of $C_6H_5N_2Cl$ in the solution after 1.00 hour.

On combining equations (2), (3), and (4) we obtain

$$n_{N_2} = c_0 V - cV$$
$$= V(c_0 - c) \qquad (5)$$

To calculate n_{N_2} we must therefore know c; this could be calculated from the rate constant given by making use of the equation

$$\log_{10} c = -\frac{kt}{2.303} + \log_{10} c_0 \qquad (6) \text{ **Equation 12.1b**}$$

c can be calculated by equation (6) and hence the required quantity n_{N_2} by equation (5).

Step 4 Collect the data, check the units, and calculate

$$k = 1.52 \times 10^{-3} \text{ s}^{-1}$$

$$t = 1.00 \text{ hour} = 3600 \text{ s}$$

$$c_0 = 1.000 \text{ mol dm}^{-3}$$

By equation (6) we therefore have

$$\log_{10} c = \frac{(1.52 \times 10^{-3} \text{ s}^{-1})(3600 \text{ s})}{2.303} + \log_{10} 1$$

from which on solving we obtain

$$c = 4.2 \times 10^{-3} \text{ mol dm}^{-3}$$

By equation (5) we then have

$$n_{N_2} = 0.0500 \text{ dm}^3 \times (1.000 - 0.0042) \text{ mol dm}^{-3}$$

$$= 0.0498 \text{ mol}$$

The moles of nitrogen evolved from the solution given in 1.00 hour is therefore 0.0498 mol.

Step 5 Review and check the solution

Example 12.3 Disintegration of radioactive nuclides

The disintegration of radioactive nuclides such as ^{238}U, ^{226}Ra, ^{90}Sr, and ^{60}Co are found to be first order. If the *half-life* for the disintegration of ^{60}Co is 5.25 years, calculate the time taken for 1.000 g of ^{60}Co to become 0.100 g. The half-life for a first order reaction is defined as the time taken for the concentration of the reactant to be reduced to half the initial concentration.

Solution

Step 1 Clarify and define the problem

We have to find the time t taken for a known amount of reaction in a first order reaction.

Step 2 Select the key equation

To find t we have to start with an equation which involves t—**Equation 12.1b**. This on rearrangement gives

$$t = \frac{2.303 \log_{10} (c_0/c)}{k} \qquad (1)$$

Step 3 Derive the equation for the calculation

To calculate t by equation (1) we require the value of c corresponding to time t and also c_0 and k.

c_0 and c can be related to the data given—the initial mass (m_0) and the mass after disintegration (m)—by the equation

$$\frac{c_0}{c} = \frac{m_0}{m} \qquad (2) \text{ (see note 1 below)}$$

In equation (1) k is also an unknown; it could, however, be calculated from the half-life ($t_{1/2}$) given. $t_{1/2}$ is the time taken to reduce the initial concentration (c_0) to half the initial concentration (i.e. to $\frac{1}{2}c_0$). This means that if in equation (1) we put $t = t_{1/2}$, then the corresponding value for c will be $\frac{1}{2}c_0$. If in equation (1) we replace t by $t_{1/2}$ and c by $\frac{1}{2}c_0$ we have

$$t_{1/2} = \frac{2.303 \log_{10} (2c_0/c_0)}{k}$$

$$\therefore \qquad k = \frac{2.303 \log_{10} 2}{t_{1/2}} \qquad (3)$$

We can now replace the unknowns (c/c_0 and k) in equation (1) by equations (2) and (3). We then obtain

$$t = \frac{2.303 \log_{10} (m_0/m)}{2.303 \log_{10} 2} \times t_{1/2}$$

$$= \frac{\log_{10} (m_0/m) \times t_{1/2}}{\log_{10} 2} \tag{4}$$

Step 4 Collect the data, check the units, and calculate

$$m_0 = 1.000 \text{ g}$$

$$m = 0.100 \text{ g}$$

$$t_{1/2} = 5.25 \text{ year}$$

$$\therefore \quad t = \frac{\log_{10} (1.00/0.10) \times 5.25}{\log_{10} 2} \text{ year}$$

$$= 17.45 \text{ year}$$

The time taken for 1.000 g of ^{60}Co to become 0.100 g is therefore 17.45 years.

Step 5 Review and check the solution

Exercise 12.3

The concentration of benzenediazonium chloride in a solution reduces to a quarter of its initial value after 80.0 minutes. Calculate the rate constant if the decomposition is first order.

Exercise 12.4

An important naturally occurring radioactive isotope used for finding the age of historical wood specimens is ^{14}C. This emits β-rays (electrons) on radioactive disintegration. A trace of this isotope is present in all samples of carbon and carbon compounds.

A sample of wood found in an ancient Egyptian tomb releases 8 electrons per minute per gram of carbon in the wood. Present-day wood releases 15 electrons per minute per gram of carbon. Calculate the age of the tomb. The half-life for the decomposition of ^{14}C is 5700 years.

Exercise 12.5

Show that, for a first order reaction, the time taken for the concentration of a substance to be reduced to half its initial value is independent of the initial concentration.

Exercise 12.6

The rate of the gaseous reaction between hydrogen (H_2) and iodine (I_2) is found experimentally to be proportional to the product of *two* concentration terms, the concentration of hydrogen and the concentration of iodine; this reaction is said to be a second order reaction.

Express the dependence of rate of reaction between hydrogen and iodine in the form of an equation.

12.2　Variation of rate constant with temperature

The rate constant k of a reaction increases with temperature T. This increase is described quantitatively by the Arrhenius equation

$$k = A \, e^{-E/RT}$$ **Equation 12.2**

where A, E, and R are constants: R is the gas constant; E is known as the activation energy for the reaction, and the pre-exponential factor A (like E) depends on the particular reaction.

Example 12.4　Activation energy from variation of rate constant with temperature

(a) Show how the activation energy of a reaction could be obtained graphically by making use of the Arrhenius equation $k = A \, e^{-E/RT}$.

(b) If the rate constant for a reaction is 1.00×10^{-3} s^{-1} at 300 K and 4.60×10^{-3} s^{-1} at 310 K, calculate the activation energy of the reaction.

Solution to Part (a)

Step 1　Clarify and define the problem

The simplest possible graph is a straight line graph. The equation for a straight line graph is $y = mx + c$ where y and x are the variables.

Step 2　The key equation

The key equation is the Arrhenius equation

$$k = A \, e^{-E/RT}$$ (1) **Equation 12.2**

Step 3　Derive the equation for the graphical plot

To obtain E graphically it would be convenient if we could first rearrange the Arrhenius equation into the form $y = mx + c$ and then plot accordingly. This could be done by taking logarithms to the base e of the Arrhenius equation $k = A \, e^{-E/RT}$ when we obtain

$$\log_e k = \log_e A + \log_e (e^{-E/RT})$$

$$= \log_e A - \frac{E}{RT}$$

$$\therefore \quad 2.303 \log_{10} k = 2.303 \log_{10} A - \frac{E}{RT}$$

$$\therefore \quad \log_{10} k = - \frac{E}{2.303R} \times \frac{1}{T} + \log_{10} A \quad (2)$$

Equation (2) is of the form $y = mx + c$. A plot of $\log_{10} k$ against $1/T$ would therefore give a straight line if Arrhenius equation is obeyed. The activation energy E can be obtained from the gradient since gradient $= -E/2.303R$.

Step 4

Does not apply.

Solution to Part (b)

Step 1 Clarify and define the problem

The activation energy E of a reaction has to be calculated. Rate constants (k) are given at two temperatures. We have therefore to derive an equation which relates E to the two k values.

Step 2 Select the key equation

We have to calculate E from k data. We therefore start with an equation which relates E and k. This is the Arrhenius equation

$$k = A \, e^{-E/RT} \qquad\qquad \text{(1) Equation 12.2}$$

Step 3 Derive the equation for the calculation

Look at equation (1). To calculate E we have to know the value of k at a temperature T and also values for the constants A and R. Out of these only the value of A is not known. A can, however, be eliminated as follows since k data are given at two temperatures.

Consider two temperatures T_1 and T_2 at which the rate constants are k_1 and k_2 respectively. At temperature T_1 equation (1) would be

$$k_1 = A \, e^{-E/RT_1} \qquad\qquad (2)$$

Similarly at temperature T_2 we obtain

$$k_2 = A \, e^{-E/RT_2} \qquad\qquad (3)$$

The unknown quantity A can now be eliminated by dividing equation (2) by equation (3) when we obtain

$$\frac{k_1}{k_2} = \frac{e^{-E/RT_1}}{e^{-E/RT_2}} \qquad\qquad (4)$$

Equation (4) can be used to calculate E since this is the only unknown.

Equation (4) is, however, inconvenient for calculation because it involves exponential terms. The equation can be simplified by taking logarithms to the base e when we obtain

$$\log_e \left(\frac{k_1}{k_2}\right) = \log_e \left(\frac{e^{-E/RT_1}}{e^{-E/RT_2}}\right)$$

$$= \log_e e^{-E/RT_1} - \log_e e^{-E/RT_2}$$

$$= -\frac{E}{RT_1} - \left(-\frac{E}{RT_2}\right)$$

$$= \frac{E}{R}\left(\frac{1}{T_2} - \frac{1}{T_1}\right) \qquad\qquad (5)$$

Equation (5) can be rewritten in terms of $\log_{10}$ as

$$2.303 \log_{10}\left(\frac{k_1}{k_2}\right) = \frac{e}{R}\left(\frac{1}{T_2} - \frac{1}{T_1}\right)$$

$$\therefore \qquad \log_{10}\left(\frac{k_1}{k_2}\right) = \frac{E}{2.303R}\left(\frac{1}{T_2} - \frac{1}{T_1}\right) \qquad (6)$$

Equation (6) can be used for the calculation of the required quantity E.

Step 4 Collect the data, check the units, and calculate

$$k_1 = 1.00 \times 10^{-3}\,s^{-1}, \quad T_1 = 300\,K$$

$$k_2 = 4.60 \times 10^{-3}\,s^{-1}, \quad T_2 = 310\,K$$

$$R = 8.314\,J\,K^{-1}\,mol^{-1}$$

$$\therefore \qquad \log_{10}\left(\frac{1.00 \times 10^{-3}}{4.60 \times 10^{-3}}\right) = \frac{E}{2.303 \times 8.314\,J\,K^{-1}\,mol^{-1}}$$

$$\times \left(\frac{1}{310\,K} - \frac{1}{300\,K}\right)$$

$$\therefore \qquad E = 1.18 \times 10^5\,J\,mol^{-1}$$

The activation energy for the reaction is therefore $1.18 \times 10^5\,J\,mol^{-1}$.

Step 5 Review and check the solution

Exercise 12.7

The activation energy of a reaction is 48.8 kJ. If the rate constant at 300 K is $1.80 \times 10^{-4}\,s^{-1}$, calculate the rate constant at 320 K

Review of Chapter 12

Only a very restricted section of chemical kinetics was considered in this chapter.

The problems considered are based on two equations—the equation for a first order reaction in its differential form (**12.1a**) and in its integrated form (**12.1b**); and the equation for the variation of rate constant with temperature.

$$-\frac{dc_A}{dt} = kc_A \qquad\qquad \textbf{12.1a}$$

$$\log_{10} c_A = -\frac{kt}{2.303} + \log_{10} c_{A0} \qquad\qquad \textbf{12.1b}$$

$$k = A\,e^{-E/RT} \qquad\qquad \textbf{12.2}$$

where $-\dfrac{dc_A}{dt}$ is the rate of decrease of the concentration of A

k is the rate constant for a first order reaction

c_A is the concentration of A at time t

t is the time

c_{A_0} is the initial concentration of A

A is a constant and depends on the reaction

E is the activation energy

R is the gas constant

T is the temperature

Appendix 1 Physical constants

Avogadro constant	L	$6.022 \times 10^{23} \text{ mol}^{-1}$
Faraday constant	F	$9.649 \times 10^4 \text{ C mol}^{-1}$
Gas constant	R	$8.314 \text{ J K}^{-1} \text{ mol}^{-1}$
Velocity of light	c	$2.998 \times 10^8 \text{ m s}^{-1}$
Freezing point depression constant for water	K_f	$1.86 \text{ K kg mol}^{-1}$
Boiling point elevation constant for water	K_p	$0.51 \text{ K kg mol}^{-1}$
Ionic product for water	K_w (298 K)	$1.00 \times 10^{-14} \text{ mol}^2 \text{ dm}^{-6}$

Appendix 2 Relative atomic masses

Element	Symbol	Atomic number	Atomic mass	Element	Symbol	Atomic number	Atomic mass
Actinium	Ac	89	227	Mercury	Hg	80	200.6
Aluminium	Al	13	27	Molybdenum	Mo	42	96
Americium	Am	95	243	Neodymium	Nd	60	144.2
Antimony	Sb	51	121.7	Neon	Ne	10	20.2
Argon	Ar	18	40	Neptunium	Np	93	237
Arsenic	As	33	75	Nickel	Ni	28	58.7
Astatine	At	85	210	Niobium	Nb	41	93
Barium	Ba	56	137.3	Nitrogen	N	7	14
Berkelium	Bk	97	249	Nobelium	No	102	254
Beryllium	Be	4	9	Osmium	Os	76	190.2
Bismuth	Bi	83	209	Oxygen	O	8	16
Boron	B	5	10.8	Palladium	Pd	46	106.4
Bromine	Br	35	80	Phosphorus	P	15	31
Cadmium	Cd	48	112.4	Platinum	Pt	78	195
Caesium	Cs	55	133	Plutonium	Pu	94	242
Calcium	Ca	20	40	Polonium	Po	84	210
Californium	Cf	98	251	Potassium	K	19	39
Carbon	C	6	12	Praseodymium	Pr	59	141
Cerium	Ce	58	140	Promethium	Pm	61	145
Chlorine	Cl	17	35.5	Protactinium	Pa	91	231
Chromium	Cr	24	52	Radium	Ra	88	226
Cobalt	Co	27	59	Radon	Rn	86	222
Copper	Cu	29	63.5	Rhenium	Re	75	186.2
Curium	Cm	96	247	Rhodium	Rh	45	103
Dysprosium	Dy	66	162.5	Rubidium	Rb	37	85.5
Einsteinium	Es	99	254	Ruthenium	Ru	44	101
Erbium	Er	68	167.3	Samarium	Sm	62	150.3
Europium	Eu	63	152	Scandium	Sc	21	45
Fermium	Fm	100	253	Selenium	Se	34	79
Fluorine	F	9	19	Silicon	Si	14	28
Francium	Fr	87	223	Silver	Ag	47	108
Gadolinium	Gd	64	157.2	Sodium	Na	11	23
Gallium	Ga	31	69.7	Strontium	Sr	38	87.6
Germanium	Ge	32	72.6	Sulphur	S	16	32
Gold	Au	79	197	Tantalum	Ta	73	181
Hafnium	Hf	72	178.5	Technetium	Tc	43	99
Helium	He	2	4	Tellurium	Te	52	127.6
Holmium	Ho	67	165	Terbium	Tb	65	159
Hydrogen	H	1	1	Thallium	Tl	81	204.4
Indium	In	49	115	Thorium	Th	90	232
Iodine	I	53	127	Thulium	Tm	69	169
Iridium	Ir	77	192.2	Tin	Sn	50	118.7
Iron	Fe	26	56	Titanium	Ti	22	48
Krypton	Kr	36	83.8	Tungsten	W	74	183.8
Lanthanum	La	57	139	Uranium	U	92	238
Lawrencium	Lw	103	257	Vanadium	V	23	51
Lead	Pb	82	207.2	Xenon	Xe	54	131.3
Lithium	Li	3	7	Ytterbium	Yb	70	173
Lutetium	Lu	71	175	Yttrium	Y	39	89
Magnesium	Mg	12	24.3	Zinc	Zn	30	65.4
Manganese	Mn	25	55	Zirconium	Zr	40	91.2
Mendelevium	Md	101	256				

Appendix 3 Defining equations

Number	Title	Equation	Physical quantities	Units*
3.1(a)	Relative atomic mass	$A_r(x) = \dfrac{\text{mass of one atom of } x}{1/12\text{th of the mass of a } ^{12}\text{C atom}}$	A_r, relative atomic mass	None (ratio)
3.1(b)	Relative molecular mass	$M_r(y) = \dfrac{\text{mass of one molecule of } y}{1/12\text{th of the mass of a } ^{12}\text{C atom}}$	M_r, relative molecular mass	None (ratio)
3.2	Avogadro constant	$L = \dfrac{N}{n}$	L, Avogadro constant N, number of particles n, amount of substance	mol^{-1} None mol
3.3	Molar mass	$M = \dfrac{m}{n}$	M, molar mass n, amount of substance m, mass	kg mol^{-1} (g mol^{-1}) mol kg (g)
4.1	% Composition	$\%B \text{ (by mass)} = \dfrac{m_B}{m} \times 100$	m_B, mass of B m, total mass of the system	kg (g) kg (g)
4.2	Concentration	$c_B = \dfrac{n_B}{V}$	c_B, concentration of B† n_B, amount, moles, of B V, volume	mol m^{-3} (mol dm^{-3}) mol $\text{m}^3 \text{ (dm}^3)$

			Quantities	Units
4.3	*Molality*	$m_B = \dfrac{n_B}{m_{solvent}}$	m_B, molality, molal concentration of B	mol kg⁻¹
			n_B, amount, moles, of B	mol
			$m_{solvent}$, mass of solvent	kg
4.4	*Mole fraction*	$x_B = \dfrac{n_B}{n}$	x_B, mole fraction	None (ratio)
			n_B, amount, moles, of B	mol
			n, total amount, moles, of all substances in the system	mol
5.1	*Ideal gas law*	$pV = nRT$	p, pressure	Pa
			V, volume	m³ (dm³, cm³)
			n, amount of substance	mol
			R, gas constant	J K⁻¹ mol⁻¹
			T, temperature	K
5.2	*Dalton's law of partial pressures*	$p = \sum p_i$	p, total pressure of system	Pa
			p_i, partial pressure of gas i	Pa
5.3	*Kinetic theory*	$p = \dfrac{Nmc_{r.m.s.}^2}{3V}$	N, total number of particles	None
			m, mass of each particle	kg (g)
			$c_{r.m.s.}$, root mean square velocity of the particles	m s⁻¹ (cm s⁻¹)
			V, volume	m³ (cm³)

* Coherent SI units are shown, but in brackets are the units most commonly used and which are based on SI. See Table 2.4 page 23 for a full list of SI units and derived units.

† This was formerly called molar concentration or molarity but these terms should be avoided. The symbol M is still sometimes used to represent concentrations (e.g. a 0.1 M solution means a solution having a concentration of 0.1 mol dm⁻³) and so is still included in some examples in this book for practice. The word litre (1 dm³) is not used. Concentration of B can also be represented by [B].

Number*	Title	Equation	Physical quantities	Units*
6.1	Raoult's law	$p_A = p_A^0 x_A$	p_A, vapour pressure due to component A in solution	Pa
			p_A^0, vapour pressure of pure A	Pa
			x_A, mole fraction of A	None
6.2	Freezing point depression	$T_f^0 - T_f = K_f m$	T_f^0, freezing point of pure solvent	K
			T_f, freezing point of solution	K
			K_f, freezing point depression constant	K kg mol^{-1}
			m, molality of solution	mol kg^{-1}
6.3	Boiling point elevation	$T_b - T_b^0 = K_b m$	T_b, boiling point of solution	K
			T_b^0, boiling point of pure solvent	K
			K_b, boiling point elevation constant	K kg mol^{-1}
			m, molality of solution	mol kg^{-1}
6.4	Osmotic pressure	$\Pi = RTc$	Π, osmotic pressure	Pa
			R, gas constant	J K^{-1} mol^{-1}
			T, temperature	K
			c, total concentration of all solute species	mol m^{-3} (mol dm^{-3})
7.1	Equilibrium constant	$K = \dfrac{\ldots [Y]^y [Z]^z}{[A]^a [B]^b \ldots}$ for the chemical equation $aA + bB \ldots \rightleftharpoons \ldots yY + zZ$	$[Y]$ etc., concentration of Y	mol m^{-3} (mol dm^{-3})
			K, equilibrium constant	depends on particular equilibrium
8.1	Logarithmic scales e.g. pH	$pX = -\log_{10} X$	X some property (e.g.$[H^+]$) pX property expressed logarithmically	$[H^+]$ in mol dm^{-3}
9.1	Internal energy	$\Delta U = q + w$	ΔU, change in internal energy of a system	J (kJ)

	Equation	Symbols	Units
		q, heat energy applied to the system	J (kJ)
		w, mechanical energy (work) supplied to the system	J (kJ)
10.1 *Electrode (redox) potential* (Nernst equation)	$E = E^{\ominus} - \dfrac{2.303RT}{zF} \log_{10} \dfrac{[X]^x[Y]^y}{[A]^a[B]^b}$ $aA + bB + ze \rightarrow xX + yY$	E, electrode potential $E^{\ominus}$, standard electrode potential R, gas constant T, temperature z, number of electrons transferred in the reaction F, Faraday constant	V V $J\,K^{-1}\,mol^{-1}$ K None $C\,mol^{-1}$
10.2 *Electromotive force of a cell*	$E = E_R - E_L$	E, e.m.f. of a cell E_R, electrode potential of the electrode written on the right E_L, electrode potential of the electrode written on the left	V V V
11.1 *Conductance*	$G = \dfrac{1}{R}$	G, conductance R, resistance	$S\,(\Omega^{-1})$ Ω
11.2 *Conductivity*	$G = \dfrac{\kappa a}{l}$	G, conductance κ, conductivity a, cross section area of electrodes l, distance between the electrodes	$S\,(\Omega^{-1})$ $S\,m^{-1}\,(\Omega^{-1}\,cm^{-1})$ $m^2\,(cm^2)$ $m\,(cm)$
11.3 *Molar conductivity*	$\Lambda = \dfrac{\kappa}{c}$ As $c \rightarrow 0, \Lambda \rightarrow \Lambda^{\infty}$ Λ^{∞} molar conductivity at infinite dilution	Λ, molar conductivity κ, conductivity c, concentration	$S\,m^2\,mol^{-1}\,(\Omega^{-1}\,cm^2\,mol^{-1})$ $S\,m^{-1}\,(\Omega^{-1}\,cm^{-1})$ $mol\,m^{-3}\,(mol\,cm^{-3})$

* Coherent SI units are shown, but in brackets are the units most commonly used and which are based on SI. See Table 2.4 page 23 for a full list of SI units and derived units.

Number	Title	Equation	Physical quantities	Units*
11.4(a)	Additivity of conductivities	$\kappa = \sum \kappa_i$	κ, conductivity of solution κ_i, conductivity due to an ion i in the solution	S m^{-1} (Ω^{-1} cm^{-1}) S m^{-1} (Ω^{-1} cm^{-1})
11.4(b)	Additivity of molar conductivities	$\Lambda = \sum \nu_i \Lambda_i$	Λ, molar conductivity of solution ν_i, number of ions of type i Λ_i, molar conductivity due to ion i	S m^2 mol^{-1} (Ω^{-1} cm^2 mol^{-1}) None S m^2 mol^{-1} (Ω^{-1} cm^2 mol^{-1})
11.5	Transference number	$t_i = \dfrac{\kappa_i}{\kappa}$	t_i transference number of an ion i κ_i, conductivity due to ion i κ, total conductivity of solution	None (ratio) S m^{-1} (Ω^{-1} cm^{-1}) S m^{-1} (Ω^{-1} cm^{-1})
12.1(a)	First order rate equation (differential form)	$-\dfrac{dc_A}{dt} = kc_A$	$-\dfrac{dc_A}{dt}$, rate of decrease in the concentration of A k, rate constant c_A, concentration of A at time t	mol m^{-3} s^{-1} (mol dm^{-3} s^{-1}) s^{-1} mol m^{-3} (mol dm^{-3})
12.1(b)	First order rate equation (integrated form)	$\log_{10} c_A = -\dfrac{kt}{2.303} + \log_{10} c_{A_0}$	c_A, concentration of species A at time t k, rate constant t, time c_{A_0}, initial concentration of A	mol m^{-3} (mol dm^{-3}) s^{-1} s mol m^{-3} (mol dm^{-3})
12.2	Arrhenius equation	$k = A\,e^{-E/RT}$	k, rate constant A, a constant for the particular reaction E, activation energy R, gas constant T, temperature	depends on reaction depends on reaction J mol^{-1} J K^{-1} mol^{-1} K

* Coherent SI units are shown, but in brackets are the units most commonly used and which are based on SI. See Table 2.4 page 23 for a full list of SI units and derived units.

Appendix 4 Answers to Exercises

Chapter 2

Exercise 2.1

(a) 10^{-7} is a positive number

(b) 10^4, 5×10^3, 10^3, 1.5, 10^0, 10^{-6}, 8×10^{-7}, 10^{-7}, 0

(c) (i) $10\,000 = 1.0000 \times 10^4$ (ii) $0.001 = 1 \times 10^{-3}$

(iii) $100^{1.5} = (10 \times 10)^{1.5} = 10^{1.5} \times 10^{1.5} = 1.0 \times 10^3$

(d) (i) $x = 4$ (ii) $x = -3$ (iii) $x = -2$

(iv) $x = 1/(1.2 \times 10^3) = 1 \times 10^{-3}/1.2 = 8.3 \times 10^{-4}$

(e) $10^{-2} \times 10^{-7} \times 10^5 = 10^{-2-7+5} = 10^{-4}$

(f) $\dfrac{10^{-2}}{5 \times 10^{-8} \times 10^3} = \dfrac{10^{-2} \times 10^8 \times 10^{-3}}{5} = \dfrac{10^3}{5} = \dfrac{10 \times 10^2}{5} = 2 \times 10^2$ ✓

(g) $\dfrac{0.0015 \times 10^{23}}{2 \times 10^{-2}} = \dfrac{15 \times 10^{-4} \times 10^{23} \times 10^2}{2} = 7.5 \times 10^{21}$

(h) 5×10^{-4} is greater than 2.5×10^{-7} by $\dfrac{5 \times 10^{-4}}{2.5 \times 10^{-7}}$ times

$= 2 \times 10^3$ times

(i) $1.000 \times 10^2 + 1.0 \times 10^{-1} = 1.000 \times 10^2 + 0.001 \times 10^2$

$= 1.001 \times 10^2$

(j) $1.00 \times 10^{-5} - 1.00 \times 10^{-7} = 1.00 \times 10^{-5} - 0.0100 \times 10^{-5}$

$= 0.99 \times 10^{-5} = 9.9 \times 10^{-6}$

(k) $(4 \times 10^4)^{1/2} = (4)^{1/2} \times (10^4)^{1/2} = 2 \times 10^2$

(l) $(2^0 \times 10^{-2})^2 = (2^0)^2 \times (10^{-2})^2 = 1 \times 10^{-4}$

(m) $(3 \times 10^{-7})^3 = (3)^3 \times (10^{-7})^3 = 27 \times 10^{-21} = 2.7 \times 10^{-20}$

(n) $(8 \times 10^{-9})^{1/3} = (8)^{1/3} \times (10^{-9})^{1/3} = 2 \times 10^{-3}$

(o) $r = (9 \times 10^{24})^{1/3} = (9)^{1/3} \times (10^{24})^{1/3} = 2.1 \times 10^8$

(p) $1\,\text{m} = 10^3\,\text{mm}$

$\therefore (1\,\text{m})^3 = (10^3\,\text{mm})^3 = (10^3)^3(\text{mm})^3 = 10^9\,\text{mm}^3$

Exercise 2.2 $V = 2.12 \times 10^{-31}\,\text{m}^3$

Exercise 2.3 $r = 3.6 \times 10^{-15}\,\text{m}$

Exercise 2.4 $1.602 \times 10^{-19}\,\text{C}$

Exercise 2.5 $4.3 \times 10^{13}\,\text{km}$

Exercise 2.6 $V = kT$

Exercise 2.7 $v = k \times \dfrac{1}{\rho^{1/2}}; \quad \dfrac{v_1}{v_2} = \left(\dfrac{\rho_2}{\rho_1}\right)^{1/2}$

Exercise 2.8	(a) 1.00×10^{-3} (b) 10.0 (c) 1.82×10^{-5} (d) 3.16×10^{-15} mol dm^{-3}
Exercise 2.9	$E = -0.279$ V
Exercise 2.10	(a) $c = 1.3 \times 10^{-2}$ mol dm^{-3} (b) $c - 13.0$ mol m^{-3}
Exercise 2.11	80 kg m^{-3}
Exercise 2.12	(a) 36.9 °C (b) $T = 309.9$ K

Chapter 3

| Exercise 3.1 | 6.022×10^{20} atoms |

| Exercise 3.2 | $N = \dfrac{mL}{M}$ |

Since mass of a molecule $= m/N$ it follows that it is equal to M/L

Exercise 3.3	(a) 10 mol (b) 6.022×10^{24} atoms
Exercise 3.4	(a) 1.01 g mol^{-1} (b) 2.02 g mol^{-1}
Exercise 3.5	(a) 1.25×10^{-2} mol (b) 7.53×10^{21} molecules
Exercise 3.6	3×10^{-23} cm^3
Exercise 3.8	(a) 0.659 mol (b) 31.56%
Exercise 3.9	(a) Al_2O_3 (b) $BeAl_2O_4$
Exercise 3.10	16 g mol^{-1}
Exercise 3.11	907 g mol^{-1}
Exercise 3.12	C_3H_8O
Exercise 3.13	N_2, 75.53; O_2, 23.47; Ar, 1.28; CO_2, 0.046; Ne, 1.3×10^{-3}; He, 7.2×10^{-5}; Kr, 3×10^{-4}; H_2, 4×10^{-6}; O_3, 8×10^{-5}; Xe, 1×10^{-5}%.

Chapter 4

Exercise 4.1	(a) 0.0100 mol (b) 0.0200 mol (c) 6.02×10^{21} molecules
Exercise 4.2	(a) NO_2, 0.32 mol; NO, 0.18 mol; O_2, 0.09 mol (b) NO_2, 14.7 g; NO, 5.4 g; O_2, 2.9 g
Exercise 4.3	0.327 kg
Exercise 4.4	54.2 g
Exercise 4.5	112 g
Exercise 4.6	1.06 g
Exercise 4.7	190.7 g
Exercise 4.8	$I_2 + 2e^- \rightarrow 2I^-$ $2S_2O_3^{2-} \rightarrow S_4O_6^{2-} + 2e^-$
Exercise 4.9	0.7332 g
Exercise 4.10	$C_8H_{18} + 12.5O_2 \rightarrow 8CO_2 + 9H_2O$
Exercise 4.11	The equation contradicts the law of conservation of charge. The two electrons (2e) should really appear on the left side of the equation.
Exercise 4.12	50.00 g

Exercise 4.13	13.4 mol dm^{-3}; dilute 74.6 cm^3 of the solution to 1.00 dm^3
Exercise 4.14	Dilute 517.3 cm^3 of solution to 1000 cm^3 with water
Exercise 4.15	0.7055 mol kg^{-1}
Exercise 4.16	79.74 g mol^{-1}

Exercise 4.17

	A	B	C
x	0.32	0.27	0.41
$10^{-3}\,c/\text{mol dm}^{-3}$	3.5	3.0	4.5

Exercise 4.18	N_2, 0.3077; H_2, 0.5385; NH_3, 0.1538
Exercise 4.19	[H$^+$] 0.032 mol dm^{-3}, [H$_2$A] 0.072 mol dm^{-3}
Exercise 4.20	0.024 mol dm^{-3}
Exercise 4.21	2.500 × 10^{-3} mol
Exercise 4.22	(a) at 25 °C, c 1.000 mol dm^{-3}; at 50 °C, c 0.971 mol dm^{-3} (b) m 1.117 mol kg^{-1} at both 25 °C and 50 °C
Exercise 4.23	4.683 × 10^{-2}
Exercise 4.24	193.2 g mol^{-1}
Exercise 4.25	(a) 44.42 cm^3 (b) 5.000 cm^3
Exercise 4.26	3.75 × 10^{-3} mol
Exercise 4.27	0.360 g
Exercise 4.28	81.82 cm^3
Exercise 4.29	(a) 3.814 g (b) 0.9221 g
Exercise 4.30	10.00 cm^3
Exercise 4.31	32.77 cm^3
Exercise 4.32	1.49 × 10^{-3} mol, 4.09%
Exercise 4.33	1.12%
Exercise 4.34	254.2 g mol^{-1}

Chapter 5

Exercise 5.3	2.4 × 10^9 molecules
Exercise 5.4	4.9 × 10^4 Pa
Exercise 5.5	(a) molar mass of the gas is required since the equation for the calculation would be

$$p = \frac{m}{MV}\,RT$$

(b) no further data are required since $\rho = \dfrac{m}{V}$ and both m and V are given in the data

Exercise 5.6	80.2 mol m^{-3}; the calculation equation would be $c = p/RT$
Exercise 5.7	44.2 g mol^{-1}
Exercise 5.8	CH_4. To answer this question you have to calculate the molar mass of the

gas; you would then get $M = 16$ g mol^{-1}. This value indicates that the gas considered is CH_4 whose molar mass, from the molecular formula, should be $(12.0 + 4 \times 1.0) = 16.0$.

Exercise 5.9	56.3 g mol^{-1}
Exercise 5.10	(a) 0.3349 (b) 3.393×10^4 Pa (c) 38.26% (d) 33.49%
Exercise 5.11	(a) 1.45×10^5 Pa (b) 1.32×10^5 Pa (c) 1.37×10^5 Pa
Exercise 5.12	To answer this question you should derive the equation which shows the relationship between $c_{\text{r.m.s.}}$ and the data given (V, n, p). This equation is $c_{\text{r.m.s.}} = (3 pV/Nm)^{1/2}$. The mass of the gas, Nm, must therefore also be known in order to calculate $c_{\text{r.m.s.}}$.
Exercise 5.13	(a) yes (b) no (c) no (d) yes
Exercise 5.14	6.29×10^2 m s^{-1}
Exercise 5.15	3.74×10^2 m s^{-1}
Exercise 5.16	(a) N_2, 0.78 mol; O_2, 0.21 mol; Ar, 9.5×10^{-3} mol; CO_2, 5×10^{-4} mol (b) N_2, 75.4%; O_2, 23.2%; Ar, 1.31%; CO_2, 0.08% (c) 1.185 kg m^{-3}
Exercise 5.17	24.0 dm^3
Exercise 5.18	2.62 g
Exercise 5.19	1.59 g
Exercise 5.20	(a) $x_{SO_2} = x_{Cl_2}$, 0.259; $x_{SO_2Cl_2}$, 0.481 (b) $p_{SO_2} = p_{Cl_2}$, 1.38×10^4 Pa; $p_{SO_2Cl_2}$, 2.56×10^4 Pa
Exercise 5.21	C_3H_8
Exercise 5.22	25.0 cm^3
Exercise 5.23	x_{AO_2}, 0.842; $x_{A_2O_3}$, 0.105; x_{O_2}, 0.053
Exercise 5.24	(a) $\rho = Mc$ where $M =$ molar mass (b) $p = \dfrac{3P}{c_{\text{r.m.s.}}^2}$
Exercise 5.25	$^{235}UF_6$ would effuse 1.0042 times faster than $^{238}UF_6$

Chapter 6

Exercise 6.1	(a) pure water (b) sucrose solution
Exercise 6.2	$P_{C_6H_5Cl}$, 1.86×10^4 Pa; $P_{C_6H_5Br}$, 1.00×10^4 Pa; P, 2.86×10^4 Pa; x, 0.650
Exercise 6.4	$$x_B = \frac{(m_B/M_B)}{(m_B/M_B) + (m_A/M_A)}$$ The equation shows that to calculate M_B from x_B, we must know the mass of solute (m_B) dissolved in a known mass of solvent (m_B), and also the molar mass of the solvent (M_A).

Exercise 6.5	189 g mol^{-1}
Exercise 6.6	(a) sugar solution (b) because the external pressure is lower at the top of a mountain than at sea level
Exercise 6.7	$0.500 \text{ mol kg}^{-1}$
Exercise 6.8	100.65 °C
Exercise 6.10	Calculation would show that the freezing point of the 30% solution is -12.8 °C. This solution is therefore unsatisfactory.
Exercise 6.11	2.00 mol kg^{-1}
Exercise 6.12	0.545
Exercise 6.13	93.5%
Exercise 6.14	(a) 100.10 °C (b) $3.29 \times 10^3 \text{ Pa}$
Exercise 6.15	$3.01 \times 10^2 \text{ mol m}^{-3}$
Exercise 6.16	$5.0 \times 10^4 \text{ g mol}^{-1}$

Chapter 7

Exercise 7.1 $1.409 \times 10^5 \text{ mol m}^{-3}$

Exercise 7.2   (a) $K_p = \dfrac{p_{NO_2}^2}{p_{N_2O_4}}$ (b) $K_p = p_{NH_3} \times p_{HCl}$ (c) $K_p = p_{CO_2}$

(d) $K_c = c_{Ag^+} \times c_{Cl^-}$ (e) $K_c = \dfrac{c_{CH_3COO^-} \times c_{H^+}}{c_{CH_3COOH}}$ (f) $K_c = c_{H^+} \times c_{OH^-}$

(g) $K_c = \dfrac{c_{Zn^{++}}}{c_{Ag^+}^2}$ (h) $K_p = p_{H_2O}$ (i) $K_c = \dfrac{c_{NH_3} \text{ (in water)}}{c_{NH_3} \text{ (in ether)}}$

Exercise 7.3 PCl_5, 0.400 mol; PCl_3, 0.100 mol; Cl_2, 0.100 mol

Exercise 7.4   α, 0.0923; % dissociation, 9.23%

Exercise 7.5   (1) $K_1 = \dfrac{c_{CO}}{c_{O_2}^{1/2}}$ (2) $K_2 = \dfrac{c_{CO_2}}{c_{CO}c_{O_2}^{1/2}}$ (3) $K_3 = \dfrac{c_{CO_2}}{c_{O_2}}$

$K_3 = K_1 \times K_2$

Exercise 7.6 K_p, $4.13 \times 10^{-11} \text{ Pa}^{-2}$; K_c, $1.46 \times 10^{-3} \text{ mol}^{-2} \text{ dm}^6$

Exercise 7.7   1.292

Exercise 7.8   3.33 atm

Exercise 7.9 4.02×10^{-4}

Exercise 7.10 $2.03 \times 10^5 \text{ Pa}$

Exercise 7.11   0.099

Exercise 7.14   (a) 0.306   (b) 0.498   (c) 0.537
(d) 0.500   (e) 0.400   (f) 0.498 mol

Exercise 7.15 $CO_2(g) \rightleftharpoons CO_2(g) + \tfrac{1}{2}O_2(g)$, mol dm^{-3}

Exercise 7.16   (a) 4.22×10^2 mol^2 dm^{-6} (b) 20.5 mol dm^{-3}

Exercise 7.17 $K_p = K_c(RT)^{-2}$

Chapter 8

Exercise 8.1   [H$^+$], 3.98×10^{-8} mol dm^{-3}
[OH$^-$], 2.51×10^{-7} mol dm^{-3}

Exercise 8.2   Calculation would show that in the solution given [OH$^-$] > [H$^+$]. The solution is therefore alkaline.

Exercise 8.3   Dilute the 11.0 M HCl solution by a factor of eleven, with distilled water.

Exercise 8.4 3.14×10^{-8} mol dm^{-3}

Exercise 8.5   5.69 mol dm^{-3}

Exercise 8.6 2.0×10^{-3}

Exercise 8.7   1.36

Exercise 8.8   [SO$_4^{2-}$], 1.23×10^{-2} M; [H$^+$], 0.162 M; [HSO$_4^-$], 0,138 M

Exercise 8.9   [H$_3$PO$_4$], 0.073 M; [H$^+$], 2.74×10^{-2} M;
[H$_2$PO$_4^-$], 2.74×10^{-2} M; [HPO$_4^{2-}$], 6.2×10^{-8} M;
[PO$_4^{3-}$], 2×10^{-18} M

Exercise 8.10 5.52×10^{-10} mol dm^{-3}

Exercise 8.12   (a) [SO$_4^{2-}$], 0.025 M; [H$^+$], 5.2×10^{-6} M;
[NH$_3$], 5.2×10^{-6} M; [NH$_4^+$], 0.050 M; [OH$^-$], 1.9×10^{-9} M
(b) 1.82×10^{-5} mol dm^{-3}

Exercise 8.13   0.0226 mol

Exercise 8.14   4.46

Exercise 8.15   15.79 cm^3

Exercise 8.16   (a) pH decrease = 0.0435   (b) pH decrease = 0.0872
(c) pH decrease = 0.477

Exercise 8.17   (a) pH decrease = 0.044   (b) pH decrease = 0.043
(c) pH decrease = 0.045   (d) pH decrease = 0.056

Exercise 8.18   (a) pH decrease = 0.176   (b) pH increase = 0.176

Exercise 8.19   (a) 2.96×10^{-14} mol^3 dm^{-9} (b) 2.96×10^{-12} mol dm^{-3}

Exercise 8.20 5.78×10^{-5} mol^2 dm^{-6}

Exercise 8.21   (a) 5.0×10^{-3} mol   (b) 1.17 g
[Cl$^-$], 5.0×10^{-2} M; [NO$_3^-$], 5.0×10^{-2} M;
[Ag$^+$], 7.8×10^{-10} M; [I$^-$], 1.9×10^{-7} M

Exercise 8.22   [A$^-$] = [H$^+$], 8.2×10^{-3} M; [HA], 9.18×10^{-2} M;
[OH$^-$], 1.22×10^{-12} M

Exercise 8.23   [H$^+$], 0.0401 M; [A^{2-}], 0.0005 M

Exercise 8.24   (a) 5.0   (b) 6.7   (c) 7.0   (d) 7.0
Remember that in an aqueous solution H$^+$ ions are produced by the dissociation of H$_2$O; this should also be taken into account. The pH of an

acid solution cannot increase to a value greater than 7.0 on dilution since $K_w = 1.0 \times 10^{-14}$ mol^2 dm^{-6}).

Exercise 8.25	0.0874 mol dm^{-3}
Exercise 8.26	12.53 cm^3

Chapter 9

Exercise 9.1	9×10^4 million kilojoules
Exercise 9.2	-1.89×10^{13} J mol^{-1}, energy is released
Exercise 9.3	(a) 3.10 kJ (b) 37.6 kJ mol^{-1}
Exercise 9.4	280 J
Exercise 9.5	-889.5 kJ mol^{-1}
Exercise 9.6	(a) endothermic (b) 1.52 kJ (c) 224 kJ
Exercise 9.7	H_2, CH_4, C_8H_{18}, C, CO
Exercise 9.9	-828 kJ mol^{-1}
Exercise 9.10	15.87 kJ; 2.45 hours
Exercise 9.11	0.724 kg
Exercise 9.12	-82.7 kJ mol^{-1}
Exercise 9.13	The standard enthalpy of formation of $CuSO_4$

Chapter 10

Exercise 10.1	(a) 2160 C (b) 4.824×10^4 J (c) 0.200 S
Exercise 10.2	(a) $C = A s$ (b) $V = kg\ m^2\ s^{-3}\ A^{-1}$ (c) $\Omega = kg\ m^2\ s^{-3}\ A^{-2}$ (d) $S = kg^{-1}\ m^{-2}\ s^3\ A^2$
Exercise 10.5	0.2235 V
Exercise 10.6	5.62×10^{-3} mol dm^{-3}
Exercise 10.7	[Cl$^-$], 0.677 mol dm^{-3}; E, 0.290 V
Exercise 10.8	(a) (i) -0.414 V (ii) 0 (iii) -0.827 V (b) (i) -0.815 V (ii) 1.23 V (iii) 0.401 V
Exercise 10.9	3.89
Exercise 10.10	-0.20V
Exercise 10.11	(a) $+1.36$ V (b) -1.36 V (c) $+0.22$ V (d) -0.49 V
Exercise 10.12	1.19 V
Exercise 10.13	1.00
Exercise 10.15	chlorine electrode, calomel electrode and silver–silver chloride electrode
Exercise 10.16	(a) Cu^{2+}/Cu (b) Pt, Br_2/Br$^-$ or Ag/AgBr, Br$^-$ (c) Pt/Fe^{3+}, Fe^{2+} electrode $-$ Fe^{2+} ions must be added to the given solution and its concentration should be known.
Exercise 10.18	7.55×10^{-5} mol dm^{-3}
Exercise 10.19	(a) 6.57×10^{-16} mol dm^{-3} (b) 6.57×10^{-16} mol^2 dm^{-6}

Exercise 10.20	Calculation shows that $E_{Ag^+/Ag} = 0.74$ V and $E_{Cu^{2+}/Cu} = 0.31$ V. Ag^+ would therefore discharge first.
Exercise 10.21	$4OH^- \rightarrow 2H_2O + O_2 + 4e$. Remember that in all aqueous solutions H^+ and OH^- ions are present and their discharge too should be considered.
Exercise 10.22	$Cu \rightarrow Cu^{2+} + 2e$. The copper in the anode would be oxidized. The data in Table 10.2 show that this reaction would take place more easily than the discharge of OH^- ions.
Exercise 10.23	894.2 C
Exercise 10.24	(a) 6.11 dm³ (b) 0.81 dm³
Exercise 10.25	964.9 C

Chapter 11

Exercise 11.1	1.25×10^{-2} S m² mol⁻¹
Exercise 11.2	3.30 V
Exercise 11.6	8.40×10^{-3} S m² mol⁻¹
Exercise 11.7	0.218
Exercise 11.8	2.73×10^{-2} S m² mol⁻¹
Exercise 11.9	20.39 S m⁻¹

Exercise 11.10

(a) $t_{(Al^{3+})} = \dfrac{2\Lambda(Al^{3+})}{2\Lambda(Al^{3+}) + 3\Lambda(SO_4^{2-})}$

(b) $t_{(SO_4^{2-})} = \dfrac{3\Lambda(SO_4^{2-})}{2\Lambda(Al^{3+}) + 3\Lambda(SO_4^{2-})}$

Exercise 11.11	0.58
Exercise 11.12	1.45×10^{-3} mol dm⁻³
Exercise 11.13	3.1×10^{-3} S m⁻¹
Exercise 11.14	1.03×10^{-14} mol² dm⁻⁶
Exercise 11.16	(a) 1.16 S m⁻¹ (b) 1.11 S m⁻¹ (c) 1.09 S m⁻¹ (d) 1.29 S m⁻¹ (e) 1.66 S m⁻¹ (f) 2.00 S m⁻¹
Exercise 11.17	(a) 1.96 S m⁻¹ (b) 1.37 S m⁻¹ (c) 1.09 S m⁻¹ (d) 1.80 S m⁻¹ (e) 3.16 S m⁻¹ (f) 4.42 S m⁻¹

Chapter 12

Exercise 12.3	2.89×10^{-4} s⁻¹
Exercise 12.4	5170 years
Exercise 12.6	$-\dfrac{dc_{H_2}}{dt} \left(\text{or} -\dfrac{dc_{I_2}}{dt} \right) = kc_{H_2}c_{I_2}$
Exercise 12.7	6.11×10^{-4} s⁻¹

Appendix 5

PERIODIC TABLE OF

1	2							
1.008 **H** Hydrogen 1 — 1								
6.939 **Li** Lithium 3 — 2)1	9.012 **Be** Beryllium 4 — 2)2							
22.990 **Na** Sodium 11 — 2)8)1	24.312 **Mg** Magnesium 12 — 2)8)2							
39.102 **K** Potassium 19 — 2)8)8)1	40.080 **Ca** Calcium 20 — 2)8)8)2	44.956 **Sc** Scandium 21 — 2)8)9)2	47.900 **Ti** Titanium 22 — 2)8)10)2	50.942 **V** Vanadium 23 — 2)8)11)2	51.996 **Cr** Chromium 24 — 2)8)13)1	54.938 **Mn** Manganese 25 — 2)8)13)2	55.847 **Fe** Iron 26 — 2)8)14)2	58.933 **Co** Cobalt 27 — 2)8)15)2
85.470 **Rb** Rubidium 37 — 2)8)18)8)1	87.620 **Sr** Strontium 38 — 2)8)18)8)2	88.905 **Y** Yitrium 39 — 2)8)18)9)2	91.220 **Zr** Zirconium 40 — 2)8)18)10)2	92.906 **Nb** Niobium 41 — 2)8)18)12)1	95.940 **Mo** Molybdenum 42 — 2)8)18)13)1	[99] **Tc** Technetium 43 — 2)8)18)14)1	101.070 **Ru** Ruthenium 44 — 2)8)18)15)1	102.905 **Rh** Rhodium 45 — 2)8)18)16)1
132.905 **Cs** Caesium 55 — 2)8)18)18)8)1	137.340 **Ba** Barium 56 — 2)8)18)18)8)2	138.910 **La** Lanthanum 57 — 2)8)18)18)9)2	178.490 **Hf** Hafnium 72 — 2)8)18)32)10)2	180.948 **Ta** Tantalum 73 — 2)8)18)32)11)2	183.850 **W** Tungsten 74 — 2)8)18)32)12)2	186.200 **Re** Rhenium 75 — 2)8)18)32)13)2	190.200 **Os** Osmium 76 — 2)8)18)32)14)2	192.200 **Ir** Iridium 77 — 2)8)18)32)15)2
[223] **Fr** Francium 87 — 2)8)18)32)18)8)1	[226] **Ra** Radium 88 — 2)8)18)32)18)8)2	227* **Ac** Actinium 89 — 2)8)18)32)18)9)2						

138.910 **La** Lanthanum 57 — 2)8)18)18)9)2	140.120 **Ce** Cerium 58 — 2)8)18)20)8)2	140.907 **Pr** Praseodymium 59 — 2)8)18)21)8)2	144.240 **Nd** Neodymium 60 — 2)8)18)22)8)2	[147] **Pm** Promethium 61 — 2)8)18)23)8)2	150.350 **Sm** Samarium 62 — 2)8)18)24)8)2
227* **Ac** Actinium 89 — 2)8)18)32)18)9)2	232.038 **Th** Thorium 90 — 2)8)18)32)18)10)2	[231] **Pa** Protactinium 91 — 2)8)18)32)20)9)2	238.030 **U** Uranium 92 — 2)8)18)32)21)9)2	[237] **Np** Neptunium 93 — 2)8)18)32)23)8)2	[242] **Pu** Plutonium 94 — 2)8)18)32)24)8)2

Key:
Atomic weight
Symbol
Name
Atomic number
Electronic structure

[ ] This is the mass number of the isotope with the longest known half life of the element indicated.
* This is the mass number of the most stable or best known isotope of the element indicated.

Index

ELEMENTS

3	4	5	6	7	0
					4.003 **He** Helium 2 — 2
10.811 **B** Boron 5 — 2)3	12.011 **C** Carbon 6 — 2)4	14.007 **N** Nitrogen 7 — 2)5	15.999 **O** Oxygen 8 — 2)6	18.998 **F** Fluorine 9 — 2)7	20.183 **Ne** Neon 10 — 2)8
26.982 **Al** Aluminium 13 — 2)8)3	28.086 **Si** Silicon 14 — 2)8)4	30.974 **P** Phosphorus 15 — 2)8)5	32.064 **S** Sulphur 16 — 2)8)6	35.453 **Cl** Chlorine 17 — 2)8)7	39.948 **Ar** Argon 18 — 2)8)8

			3	4	5	6	7	0
58.710 **Ni** Nickel 28 — 2)8)16)2	63.540 **Cu** Copper 29 — 2)8)18)1	65.370 **Zn** Zinc 30 — 2)8)18)2	69.720 **Ga** Gallium 31 — 2)8)18)3	72.590 **Ge** Germanium 32 — 2)8)18)4	74.922 **As** Arsenic 33 — 2)8)18)5	78.960 **Se** Selenium 34 — 2)8)18)6	79.909 **Br** Bromine 35 — 2)8)18)7	83.800 **Kr** Krypton 36 — 2)8)18)8
106.400 **Pd** Palladium 46 — 2)8)18)18	107.870 **Ag** Silver 47 — 2)8)18)18)1	112.400 **Cd** Cadmium 48 — 2)8)18)18)2	114.820 **In** Indium 49 — 2)8)18)18)3	118.690 **Sn** Tin 50 — 2)8)18)18)4	121.750 **Sb** Antimony 51 — 2)8)18)18)5	127.600 **Te** Tellurium 52 — 2)8)18)18)6	126.904 **I** Iodine 53 — 2)8)18)18)7	131.300 **Xe** Xenon 54 — 2)8)18)18)8
195.090 **Pt** Platinum 78 — 2)8)18)32)17)1	196.967 **Au** Gold 79 — 2)8)18)32)18)1	200.590 **Hg** Mercury 80 — 2)8)18)32)18)2	204.370 **Tl** Thallium 81 — 2)8)18)32)18)3	207.190 **Pb** Lead 82 — 2)8)18)32)18)4	208.980 **Bi** Bismuth 83 — 2)8)18)32)18)5	[210] **Po** Polonium 84 — 2)8)18)32)18)6	[210] **At** Astatine 85 — 2)8)18)32)18)7	[222] **Rn** Radon 86 — 2)8)18)32)18)8

151.960 **Eu** Europium 63 — 2)8)18)25)8)2	157.250 **Gd** Gadolinium 64 — 2)8)18)25)9)2	158.924 **Tb** Terbium 65 — 2)8)18)27)8)2	162.500 **Dy** Dysprosium 66 — 2)8)18)28)8)2	164.930 **Ho** Holmium 67 — 2)8)18)29)8)2	167.260 **Er** Erbium 68 — 2)8)18)30)8)2	168.934 **Tm** Thulium 69 — 2)8)18)31)8)2	173.040 **Yb** Ytterbium 70 — 2)8)18)32)8)2	174.970 **Lu** Lutetium 71 — 2)8)18)32)9)2
[243] **Am** Americium 95 — 2)8)18)32)25)8)2	[247] **Cm** Curium 96 — 2)8)18)32)25)9)2	[249] **Bk** Berkelium 97 — 2)8)18)32)27)8)2	[251] **Cf** Californium 98 — 2)8)18)32)28)8)2	[254] **Es** Einsteinium 99 — 2)8)18)32)29)8)2	[253] **Fm** Fermium 100 — 2)8)18)32)30)8)2	[256] **Md** Mendelevium 101 — 2)8)18)32)31)8)2	254* **No** Nobelium 102 — 2)8)18)32)32)8)2	[257] **Lw** Lawrencium 103 — 2)8)18)32)32)9)2